Lessons from pond scum

The Eco-jar cautions against our technological view of Earth's Life System

Ted Merrill

Lessons from pond scum
by Ted Merrill

ISBN 978-0-9773348-3-4

Published by Homeostasis Press
5919 147th St. SW, Edmonds WA 98026

Contents

Author's dedication

This book is dedicated to Bill Merrill, my brother and lifelong friend. Since childhood we continually exchanged, challenged, fertilized, and enhanced each other's ideas and world views, even though usually from an extended geographical distance. After entering our ninth decade, we recognized that we needed to acknowledge our impending mortality as an integral part of the grand dynamics of the ecosphere.

Bill was first to reach the finish line of his life, during the writing of this book. But his own book, *Wisdom of the Tools* (also from Homeostasis Press), has been a key contributor of history and insight to these pages.

About the author

I am one of two sons of schoolteacher parents, and I've lived most of my years in small towns in Idaho, California, Oregon, and Vermont. I feel blessed to have had my childhood during the Great Depression without ever feeling that we were poor or disadvantaged in any way, and further blessed to have reached maturity before the advent of plastics, television, computers and calculators, video games, cell phones, the internet, the iPod and Skype and Twitter and Facebook and whatever may come next.

I am more than a decade retired from 50 years practicing medicine, alternating between rural general practice and small city emergency departments, and in mid-career took a 2-year sabbatical to teach at Goddard College, a wonderfully weird little place in Vermont.

Since childhood my relationship with the ecosystem, besides my being an integral and grateful part of it, has included working and playing in gardens and fields, rivers and mountains, laboratories and classrooms, fishing and hunting, peering through a microscope, and assisting fellow humans where needed in the awesome processes of birth, living, and dying. And now more than ever I am focused on my identity as an integral part of the natural world, an ethical human organism with that unique property of humans, the churning of questions and ideas into some sort of integrated sense of how the world works.

Lest my arguments, beliefs, and motivations be misunderstood, I would like to clarify my position yet further. I am an American. I was born in the United States, have lived and grown old here, and will die here. Among all the places on Earth that I have seen, there is none in which I would prefer to live. Even on crossing the border when returning from a visit to Canada, I have a relieved feeling of having returned safely home.

I have served in both the U.S. Navy and the U.S. Army, though chance circumstances kept me out of combat: I was never shot at, pursued, nor in real personal danger. I have voted in every election since my age permitted it, and from time to time I communicate with my elected representatives to exert whatever minor effect one voice may have. I sincerely pledge allegiance to the flag and to the Republic for which it stands. After each election I am deeply grateful for the privilege of living under the constitutional framework of our form of government, and I'm proud to live in a nation where, despite widespread and raucous political differences during campaigns and after the voting, the inauguration of new officials — the actual regime change — is carried out courteously, respectfully, almost reverently (in contrast to the violence seen in many places in the world) and the losers accept the wait for another two or four years for a chance to compete again, while the winners resume the contentious and often heated business of the political system.

The reason for this book

As my remaining days count down, I feel ever more passionately the need to share my reasons for great concern — almost desperation — about the continuation of the Life System as I have known it, and specifically about the long-term survival of the human species, on this beautiful blue-green ball whirling through space.

I was with my wife when she died of heart disease. Gradually, breath by breath, her effort became weaker and shallower, until the last life force drained away and her lips, then her lungs, and finally her heart fell silent. I saw my aged mother pass in much the same way: consciousness gone, pulse undetectable. And in both events there was a certain moment, a subtle change in the breathing, when I knew the downward spiral of vital forces had at last become irreversible.

Sometimes a chill lies on my heart as I look at our world — our "civilized" world of "modern" culture and technology — and fancy that I am seeing once more that impending change in dynamics, a shift in the delicate balance of life processes that sustain each other's flow around and through the organism that is our world. At such times, I imagine the human species having passed some invisible stage beyond rescue and drawing its final, slowly fading breaths.

From such a fantasy I look around and jerk myself back to the safety of reassurance. The reality of this present day is that the earth still turns in the sun; the breeze still sweeps away the dust and somehow returns sweet and fresh from the sea; the water pours pure and good from the mountain, and a carpet of green spreads over the valley. I can smell breakfast cooking and hear children's laughter in the street. And with the eternal optimism that is the foremost tool of the physician, I truly believe there is hope for us yet. We need only become aware of all the little details enough to show us the grand generalization, which should allow us to set ourselves in accord with the web of life.

The content of this book originated more than 40 years ago when I first began writing about homeostasis as a logical extrapolation from the Eco-

"Where Do We Come From? What Are We? Where Are We Going?"
Paul Gaugin, 1897-98

Jar (see Chapter 1) to the Earth itself. However, as it has taken shape, I've become increasingly aware that the book also revolves around Paul Gauguin's Three Great Questions. Gauguin, an eccentric (mad, some would say) French painter in the latter part of the 19th century, was obsessed through much of his life with three questions: *Where do we come from? What are we? Where are we going?* The 1897 painting he considered to be his masterpiece (shown above) dealt with those three questions, and they lurk beneath and hover over all that follows here.

Ultimately, though, this book is an extension of my medical practice.

In my long medical career, when confronted by a person in distress, it has been my chosen duty to discover the cause of the distress, to understand and explain the pathology that causes it, and to eliminate or mitigate it to the best of my ability.

The industrialized world, of which the U.S. has been the leader, has developed a distress bordering on derangement, a deep and possibly fatal pathology. It is to that pathology, and in that healing spirit, that this book is addressed.

Introduction: Earth's layer of life

The theme of this book is Life. That statement is somewhat meaningless without some context, and here is the context in which I view it.

Growing evidence says that more than 4 billion years ago, all matter in our solar system already existed in roughly the same chemical and physical form as it does now, condensed and gradually aggregated from a giant molecular cloud. And in that solar system, during all those billions of years, Life has been in the process of creating itself.

Today, our solar system consists of our home star (the sun) and the objects gravitationally bound in orbit around it. Those objects consist mainly of what are now eight planets, the closest four — Mercury, Venus, Earth, and Mars — being the terrestrial planets, consisting mostly of solid and liquid matter rather than gases. The next of the orbiting planets, Jupiter and Saturn, are "gas giants," and of those beyond, Uranus is mainly ices of water, ammonia, and methane, and Neptune is mainly hydrogen and helium.

We stand on the cooled, hardened outer crust of this "third rock from the sun," floating atop the molten stone that comprises most of the planet's insides. In, on, and above this floating crust, Life has created and now continues to recreate itself, reorganizing to survive and accommodate local conditions. It's been doing this for a couple of billion years.

As the sentient form of life on our planet, we humans continue to search for signs of life, past or present, on the other terrestrial planets, especially Mars. But none comes close to providing the life-giving qualities of our own.

Photographs taken by U.S. astronauts during the 1969 moon mission showed Earth as a beautiful blue-green ball whirling through the vacuum of space. And what is the meaning of the blue-green color? Unlike any of the other planets, ours is covered by what we might consider a membrane of living organisms.

Clearly, life exists on the surface of Earth where we live, but the essence of life extends above and below what we see on the globe. That layer of life covers the entire depth from the lowest point on the Earth's surface —the bottom of the Mariana Trench, about 36,000 feet beneath the surface of the Pacific ocean — upward to the height of the highest clouds,[1] about 60,000 feet.

That's a layer of life about 18 miles thick. At the scale of a traditional classroom globe perhaps two feet in diameter, the thickness of this layer of life is about equal that of a heavy coat of paint.

As you read this book, I encourage you to think of Earth in that way, as a ball of molten rock with a cooled, hardened surface, some 8,000 miles in diameter, turning slowly in the vastness of black space, on which exists a blue-green layer of life just 18 miles thick.

Conditions

The "Miracle Whip Microcosm"

In 1972 I quit teaching biology in Vermont and was decommissioning a salt-water aquarium. I hated to throw out its contents, including little creatures I had brought from the Maine coastal tide pools. As an experiment, I washed out a Miracle Whip jar, scooped some sand and mud from the bottom of the tank into the jar, filled it almost completely with the water, and added some algae and several of those tiny creatures. I melted some paraffin into the lid, screwed it down airtight, and set it on the windowsill.

I knew the plants (algae) would provide oxygen and food for the animals and the animals would provide the CO_2 and fertilizer needed by the plants, and with a little luck, the system would balance itself out.

The jar's appearance kept changing for several weeks. Most (but not all) of the animals died; the large leaf-like pieces of algae slowly deteriorated and were replaced by a velvety carpet of a different green on the sand and the sides of the glass. When I sold the house 3 years later and left the jar on the windowsill for the new owner, one snail still grazed his little trails through the layer of green.

Because my experiment began with a Miracle Whip jar, it seemed natural to refer to it as a "Miracle Whip Microcosm." That became everyone's favorite name for these ongoing experiments, and the Kraft Foods company has given me permission to use it in this book.[2] But all subsequent jars have been from different products, and for me, "Eco-jar" slips more easily off the tongue, so the terms are used interchangeably in these pages.

Since making that first jar in 1972, I have never been without one or more of them on my windowsill, using material dipped out of tide pools or some stagnant pond or swamp, or scraped from the pilings under the dock at a marina. They have survived from a few months to as long as 12 years, sealed airtight and operating only on the energy from sunlight.

Earth as an Eco-jar

The Earth is a spaceship in the largest sense of the word, a closed ecological system supporting a hugely diverse variety of life forms, all interdependent and self-balancing, and driven entirely by energy from the sun. Except for its scale, of both size and diversity, the Eco-jar is an authentic model of the Earth, and it is small enough to fit more easily into our minds.

By careful comparison, we can confirm its similarity to the Earth's system. We may not usually think of our planet in this way, but the similarity is undeniable when Earth is viewed from a distance.

- It is a closed ecosystem — closed except for the input of energy in the form of light radiated in, and of loss of energy in the form of heat radiated out.
- What's in there is all there is. No new supplies can be ordered in, and no garbage can be gotten rid of. What is here can only be converted to a different form or moved to a different place.
- The numbers and kinds of living organisms are regulated entirely by their own inherent properties. Those properties determine organisms' interactions with each other and with the non-living parts of the system — air, water, rocks, and movements of things by energy transfer.
- There is no free lunch. Everything that happens is caused by, and affects, everything else.
- The scale and rate of happenings cover a vast range, from the speed of light to the gradual dissolving or erosion of rocks by water.

The scientific-industrial mindset is that the Earth's life system presents many options, and science and technology are usually needed to make things work out right (that is, to the advantage of humanity). But the Eco-jar reminds us that evolution has already taken care of things, and continues to take care of them, and our intervention only limits the possible outcomes.

By thoughtful observation of Eco-jars, you can learn endless lessons, not only in physics, chemistry, biology, and ecology, but also in economics, politics, demographics, population dynamics, and more.

No two creatures are identical, which means every creature is born with a difference. Some differences render a creature less compatible with its surroundings, and some give it special survival advantages. It is these interactions not only among creatures, but also among internal and external biological subsystems, that make evolution a selective process.

Appendix A is a kind of owner's manual, with tips on how to set up an Eco-jar and avoid some of the disappointments I've experienced and mistakes I've made.

Useful concepts

The nature of the Eco-jar is one of many insights and metaphors I have found useful in thinking about how the world works and understanding why we behave as we do. To set the stage for the discussions that follow, I'd like to introduce a few more of those concepts.

How people learn

We learn by experience — a continuous series of experiences beginning before we leave the womb, and continuing until this very moment. At some indefinable time in the womb we begin to have sensory experiences of ourselves — a hand touching a leg, or possibly learning how to put a finger in our mouth — and of our environment, a limited space for movement, being tipped this way and that as measured by the orienting axis of gravity, the sound of mother's heartbeat, and meaningless sounds from outside.

At birth we encounter an avalanche of new experiences. All our senses come into play, a wealth of new stimuli, and we gradually make some connections between them — between mother's arms and the feel and taste of milk or of a clean, dry diaper. Then we begin to increase movement, to turn over, to crawl, then to walk and fall over and get up and walk again.

Learning is three-dimensional, visceral, tactile, sensory, but also mental, integrating and always assigning meaning to the sensory input. We constantly and subconsciously suppress most of that input in order to pay attention to what concerns us at the moment. Hence no two people can ever have exactly the same experiences, in part because, even if they are standing side by side and witness the same car crash, each will suppress and attend to different details.

Each new experience or observation is automatically tested, like trying the fit of a jigsaw puzzle piece, against our existing worldview to see whether it is consistent with the picture we've developed thus far. If it is, then our worldview is expanded and enriched. Otherwise, we may set the piece aside to try again later, or we may decide it is part of an entirely different puzzle and discard it completely.

Thus learning happens entirely through experiential contexts. The experience of learning about, say, an orange or a fish by viewing it on a two-dimensional photograph or computer screen would differ enormously from having it on the table in front of you, touching it, smelling and tasting it, lifting it, sensing its weight and texture. It is active rather than passive experience. Someone who grows up on a farm will have vastly different experiential contexts from someone who is a city dweller, so their views of soil, food, water, wastes, weather, winter, and economics will differ quite a bit.

When we experience verbal, written, and graphically transmitted information (this book, for example), testing the fit of the puzzle pieces can be more challenging. We are constantly bombarded with conflicting claims or evidence, and we must judge which are most credible, then add them at least tentatively to the worldview we are building.

Individual Worlds

It must follow, from the above description of learning, that each person inhabits his or her own exclusive, unique world, a world that is largely inaccessible to other persons and can be shared only partially and superficially. No matter how we may try to share our world with someone else, communication is never fully up to the task, but is rather a cause of many painful failures of relationships that, early on, were taken for granted.

However, those parts of our worldview that we can partially share, or that intersect with the worlds of others, become part of the myths and traditions that glue our culture together.

The word "myth" is often used in a pejorative or dismissive sense, claiming something to be false or totally fabricated. But the word has a more long-standing anthropological meaning, defined here by Ronald Wright:[3]

"Myth is an arrangement of the past, whether real or imagined, in patterns that reinforce a culture's deepest values and aspirations. ... Myths are so fraught with meaning that we live and die by them. They are the maps by which cultures navigate through time."

The conflict of Me

Each of us is born with a built-in conflict of interest. It is the conflict between *Me as myself* and *Me as one of Us*. This dilemma conditions almost our every act and decision, every day and throughout life. Should I take the last cookie on the plate? Who should go through the door first? Should I vote, or join a committee, or try to rescue a drowning person?

There are so many groupings of Us: concentric, meaning embedded one within the next larger group (family, neighborhood, town, state, country); intersecting, meaning not embedded but cutting across concentric lines (skills, educational past, professions, organizations, politics); relevant to the context of this book, the whole human species; and beyond that, ultimately concentric or including all of Us, the biosphere's living organisms.

The more you know, the more you don't know

I vividly remember a parable told by a visiting school speaker when I was about 14 years old:

If you sit outside at night by a small campfire, you see a small circle of light surrounded by a small circle of darkness. Add more fuel, and the fire casts a larger circle of light, surrounded by a larger circle of darkness.

My interpretation now: Each time you find the answer to a question, it unveils more questions. As your accumulated knowledge grows, the surrounding circle of questions, of mysteries, grows larger, deeper, richer — a depth and expanse of mystery without limits.

Acceptance

Don't let your attitude get bent out of shape because rocks are hard and water is wet; it could ruin your life. Accept things — and people — as they are, and then work around it.

As Lao Tsu said, *"It is the empty space inside a vessel that gives it value."*

What is ecology?

Leave no stone unreturned

Our father taught my brother and me many of life's lessons through fishing. At ages perhaps 6 and 8, we waded out with him into a shallow stream to turn over flat rocks and look for periwinkles. These hollow cases, about the size of a date seed and made of sand grains cemented together, project like fingers from the bottom of the rock. Pull one off and break it open and you will find a tough yellow wormlike creature which we know is highly attractive to trout.

We put several in our fish sacks, and Dad said, "Always take just what you need for the day's fishing. If you take a lot, so they just go to waste, sometime you may come and find there aren't any left."

Then he turned the rock back over and placed it carefully where he had found it. "That's home to other little critters besides periwinkles," he said. (We had in fact seen a hellgrammite scramble off the rock and drop into the water, and some other tiny slithery or slimy things clinging there.) "You should always put it back."

I'm quite certain that Dad never heard the word "ecology." He didn't need to.

Two opposing views of ecology

The popular view in Western cultures is the ***conservation, anti-pollution view,*** which holds that our environment is endangered because of careless or imperfect technology, and the solutions involve more technology. We must develop anti-pollution methods and greatly increase spending to combat the corruption of our environment. Inevitable exhaustion of non-renewable resources such as oil and metals are also considered technological challenges to substitute something else or somehow create replacements. In this view, environmental issues are considered "externalities" and unforeseen consequences.

The spaceship mentality, or the ***profound-change-of-values view,*** questions whether the human species can survive with the life style, value system, and relationship to each other and to the Earth that Western society is leading and much of the world is progressively following.

In this book, somewhat against the rush of popular opinion, I speak for the second view, and I argue that humankind has both the urgent need and — we may tentatively hope — the capacity to move toward a sustainable ecological philosophy before it is too late.

Perspectives of ecology

Here are some of the aphorisms of ecology, most of which are reminiscent of the "Lessons from the Eco-jar" listed in Chapter 1.

- ***The earth is a spaceship.*** There is a finite amount of material and of room here, and our problem is to learn how we can use the stuff and the room so that enough of both of them will continue to be available to us and enable us and future generations to survive.

- ***There is no place anything can be thrown away.*** It can only be moved from one place to another.

- ***All power pollutes.*** That is, possession and application of power always produces changes which, at some level, will become destructive.

- ***Nothing can be produced or consumed,*** but only converted from one form to another. Water, CO_2, and nutrients in the soil are converted by plants into groceries. Some of those are converted by animals into other types of groceries. Both are converted by us into human meat and bone, energy, and sewage. The sewage is converted by bacteria in treatment plants into liquid nutrients, which are flushed by rivers into the ocean, and into solid nutrients to be buried in garbage dumps. We convert iron ore into a car, convert the new car into an old car, and convert the old car into a part of a pile of old cars which we hide behind a high fence. We move petroleum out of the earth and convert it into asphalt roads and vehicle fuel and petrochemicals. We convert forests first into orchards or fields, then into parking lots.

- ***We are converting natural resources into artifacts faster than the processes of nature can convert them back into resources.***

Growing power, shrinking awareness

Our measure of a person is how big a ship he can sail, or how many houses he can build, or how much money he can acquire, or how fast he can go. ***All those things require power.*** One person acting alone can lift a fairly big rock, or build a house (if it's the right kind), or spade up a fair-sized piece of ground in a month. But in order to use his full capacity for creativeness, that person needs a horse to pull the plow; or better yet, a tractor and a power saw. A bigger tractor would allow him to do even more, and so on.

People who build log cabins or adobe houses know what they are doing to the environment. They see the stumps that are left, or the hole in the clay hillside. But most people who build modern houses no longer see what they are doing to the Earth. I imagine few builders stop, while sawing a board or raising a wall, to visualize the logged-off patch in an Idaho forest that this house represents, or to wonder about the environmental implications of the iron nails, the aluminum roofing, the ceramic tiles, the glass panes, or the fuel that was burned to make and transport all those things.

Clearly, some of the changes that have taken place since Eden are a great ***increase in the amount of power*** (physical energy) that can be harnessed by a person and exerted against the earth; a vast ***increase in the distant extension of a person's effect*** on the earth; and a corresponding ***decrease in awareness*** of the relationship to the earth. In a former day, in an agrarian society, people had an intimate and intense awareness of their dependence on the earth for food and security. Now that dependence is no less complete, but our sense of it has changed. We do not now feel directly dependent on soil fertility and favorable growing weather for our food, but rather on the reliability of the nearest grocery store. We've lost our connection to the fact that someone, somewhere, can only put those tomato seeds in the ground, add water, and wait.

We are not evil people for doing those things. We do them not because of something a serpent told Eve in the Garden of Eden, but because somehow the pint of magic pudding in the skull of the ape evolved into a quart of magic pudding in the skull of the human, and we are therefore able to conceive and make and use tools, and to dream and to worry, and to symbolize and pass on in language the things that happen in our heads.

But now, after going about our business for quite a long time, we're learning that some of the things we've been doing are bad for us, and it is not always easy to figure out which things they are. It is much easier to look for someone to blame for the ill effects, and to continue doing the same things, because we know (for we have always been told) that they are The Good Things To Do, and they have always seemed to work up until now.

For some 2 billion years, our Life System — the only one we know to exist in the entire universe — has maintained itself and evolved into ever-increasing complexity. But the human brain and mind have unwittingly begun to disrupt the ecological balance, continually increasing the power we can exert against the Life System, widening the reach of our effects on that System, and thus dulling our awareness of our devastating effects on that System, and also of our dependence on it. Most disturbing is that we're losing our awareness that **our dependence on the Life System is exactly as complete as that of all other creatures.**

Our perceptions of the world are limited by our life span — that is, how long each of us has already lived. We know only what we have seen: all else is hearsay. We are born into a certain world, and we must assume this is how the world is, so we learn to adjust to and function within the world as we find it. But if we live long enough, and if we pay attention, we discover that the world is rapidly changing, shifting under our feet.[4]

What I am suggesting is that **the meaning of ecology raises some painful and profound questions to which no answers are now available.** We customarily bypass these questions by thinking in a time frame short enough so the questions lose their apparent urgency. But ecology is far more than just cleaning up the roadside. It calls for a totally new order of decision making, with a shift in assumptions and of ethos, and with a new view of humanity, of the world, and especially of time.

The End Times (both ends)

They said I began in the back seat of
a Model-T on their wedding night. But no.
They each brought life from long before,
passed down, passed down and down 'til
I can almost remember the Big Bang.
— Ted Merrill

This chapter addresses the two of Gauguin's questions that bracket the whole of human existence: *Where do we come from?* and *Where are we going?* Most religions have answers to both of these questions, but because there were no eyewitnesses to our beginnings, and because I don't know the answers, your judgment on this is as good as any. I will, however, offer an explanation of my beginnings that works for me, seems the most internally consistent, and is an integration of several different views of the issue, including some derived from the biblical stories taught to me in childhood.

Practically every religion has a creation story, and each indigenous and pre-literate culture has a mythology which explains — in terms consistent with their homeland and their concept of their relationship to it — how their people came into being.

One thing needs to be established about the time involved in creation, however it happened. It is said in the Bible that in God's eyes a day is as a thousand years, and a thousand years is as a single day. So whether creation took six days, or six thousand years, or two billion years, or none of the above, cannot be stated with any certainty. However, the answer does not make any real difference to the nature of the world we presently find around us.

Emergence

A long established belief and assumption in science is that something cannot be made from nothing. This has occasionally been questioned (maggots appearing in a carcass, or bacteria in food), but ultimately these doubts have been firmly laid to rest.

Even in Genesis, the Bible story of creation, God may have had some raw materials to work with (in this version, water and "dry"):[5]

When Elohîm began to create the sky and the earth, the earth was shapeless and empty and darkness across the abyss, and Elohîm's wind swept across the waters. And Elohîm said, "Let light be." And light was. And Elohîm saw the light, that it was good. And Elohîm made a division between light and dark. And Elohîm called the light Day, and the dark he called Night. And it was evening and it was morning day one.

And Elohîm said, "Let a bowlshape be in the middle of the waters, and let it make a division between waters and waters." And Elohîm made the bowlshape, and it made a division between the waters that were underneath the bowlshape and the waters that were above the bowlshape. And it was so. And Elohîm called the bowlshape Sky. And it was evening and it was morning a second day.

And Elohîm said, "Let the water underneath Sky be gathered into one place, and let the dry appear." And it was so. And Elohîm called the dry Earth, and the gathered water he called Sea. And Elohîm saw that it was good.

Theoretical physicists have faced the same limitations of explanation. The "Big Bang" theory assumes (more or less) that everything now existing — the entire universe — was once crammed into an extremely small and hot mass that somehow exploded and expanded in the blink of an eye into atomic or sub-atomic particles. But how did it all get so severely compressed? And where did it come from?

Because it is pointless to speculate on events before the Big Bang, I'll let Harold Morowitz[6] pick up the story from there. He describes how, in

28 somewhat arbitrary steps, the elementary particles gathered together into new combinations, and each new emergence leads to the next level of possibilities.

Assume that the particles have coalesced into the atoms of the known natural elements. Atoms of hydrogen and atoms of oxygen each have their own separate, unique, and independent characteristics. If they come into contact with each other under the right conditions they will cling together to form a molecule of water. The properties of water are totally different from those of either oxygen or hydrogen, and thus the possibilities for other interactions are greatly expanded. We call that phenomenon emergence.

Emergence is enabled by the principle of homeostasis. If the new interactions of water with other things lead to combinations that include homeostatic feedback, a new "system" comes into being. Without a functional set of feedback relationships, that particular almost-system remains forever uncreated.

When the chemicals diffusing in the primordial waters came together to form the first living cell, that was emergence. When the activities of the neurons in the brain resulted in mind, that too was emergence. Morowitz illuminates the emergence of living cells, animals, vertebrates, and mammals, leading to the great apes and the appearance of humanity. He also examines tool making, the evolution of language, the invention of agriculture and technology, and the birth of cities. Emergence is a fascinating way to look at the universe and the natural world.

How long is our future?

The other end of existence — Gauguin's other question, *Where are we going?* — is just as obscure. Some religions include the expectation of a Messiah coming to end the world in one way or another. I have no credible information either way regarding this possibility, so in my ignorance

I prefer to believe there is no end in sight. It appears to me that antiquity and eternity are mirror images of each other, and that we are standing always at the exact center of time. This moment is all we have. All the rest, past and future, exist only in our minds. The past is fabricated from memory and myth, and the future is imagined extrapolations from that questionable past.

This brings us face to face with the question: **How far into the future do we care?** Just past our children? Just past *their* children? Or all the way out until the sun goes cold?

This is a crucial question, because **the answer determines whether or not we already concede that human nature is the ultimate vandal, destined to destroy the Life System and declare it a failed experiment.**

The most intriguing answer I have found is presented on a must-see web site, *thelongnow.org*. This group of people has conceived and begun to implement the idea of a 10,000-year clock to be built in a hollowed-out mountaintop in a remote area of Nevada. It will be 60 feet tall, driven by the seasonal temperature changes. It will tick once each year, chime once per century, and the cuckoo will come out at the end of each millennium. There is already an 8-foot working prototype in a Science Museum in England.

Michael Chabon, writing in January 2006,[7] points out that:

Even if the Clock of the Long Now fails to last ten thousand years, even if it breaks down after half or a quarter or a tenth that span, this mad contraption will already have long since fulfilled its purpose. Indeed the Clock may have accomplished its greatest task before it is ever finished, perhaps without ever being built at all. The point of the Clock of the Long Now is not to measure out the passage, into their unknown future, of the race of creatures that built it. The point of the Clock is to revive and restore the whole idea of the Future, to get us thinking about the Future again, to the degree if not in quite the same way that we used to do, and to reintroduce the notion that we don't just bequeath the future — though we do, whether we think about it or not. We also, in the very broadest sense of the first person plural pronoun, inherit it. ... The future, by definition, does not exist. "The Future," whether you capitalize it or not, is always just an idea, a proposal, a scenario, a sketch for a mad contraption that may or may not work. "The Future" is a story we tell, a narrative of hope, dread or wonder. And it's a story that, for a while now, we've been pretty much living without.

Hope, dread, or wonder. How familiar! These, rather than despair and defeat, are the best we have to offer our children, and our best equipment for our precarious journey toward the future. And my fondest hope is that our offspring, and theirs after them, will pick up the story with hope and wonder, and courage to match the dread, and will carry on with the grand and mysterious work of human evolution.

The simple chemical dance of life

The natural world contains more than 100 different chemical elements. Of those, carbon, hydrogen, and oxygen, with smaller amounts of nitrogen, phosphorus, and sulfur, and mere trace amounts of a dozen other elements, comprise the total chemical structure of life. Of those, carbon is the basic source of both mechanical and biological energy. That's why we sometimes refer to ourselves as "carbon-based life-forms."

The smallest unit of any element is an atom, and atoms hook up together to form molecules. For example, two atoms of hydrogen and one atom of oxygen join to form a molecule of water, indicated as H_2O. One atom of carbon and two atoms of oxygen join to form a molecule of carbon dioxide, or CO_2. By themselves, floating in the atmosphere, oxygen atoms tend to cling together in pairs, so oxygen is designated as O_2.

When you think of biochemistry, you may think of complex science requiring a college degree. But the essential chemistry of Earth's Life System is beautifully simple and well within the realm of general understanding. All that's required is a little bit of open-minded concentration.

Carbon and fuel

Carbon and its interactions with hydrogen and oxygen provide the fuel for life processes in all living things, plants and animals alike. The word "fuel" implies burning, or combining with oxygen (oxidation), which releases energy. Interestingly, those same carbon-hydrogen-oxygen interactions and energy-releasing oxidation processes operate in other carbon-based fuels as well. And because the process of burning coal, gasoline, or wood is more simple and direct, let's start there.

In its fuel function, carbon comes in three forms:

- ***Carbon alone***: Coal, charcoal.
- ***Carbon with hydrogen*** (hydrocarbons): Gasoline, other petroleum products, natural gas (methane), propane.
- ***Carbon with hydrogen and oxygen*** (carbohydrates): Cellulose (wood fiber), starches, sugars, alcohols.

The air around us is a mixture of gases, about one fifth of it being oxygen (O_2). In a coal-fired steam engine or in my charcoal barbecue, once the fuel is ignited to start the process, air flows past the fuel and provides oxygen, promoting burning (oxidation) and the release of heat. When the oxygen from the atmosphere (O_2) combines with the carbon in the fuel (C) as it burns, it creates an equivalent amount of carbon dioxide (CO_2) that is released back into the atmosphere. When wood or gasoline burns, its hydrogen also combines with oxygen to produce H_2O (water), which flies away as water vapor into the atmosphere (or drips from the tailpipe of your car).

When wood, leaves, or other plant material decays on the ground, the same conversion process happens, though much more slowly. A dead tree may take almost as long to rot away completely as it took to grow. Ultimately, though, ***exactly the same amount of heat, CO_2, and H_2O will be produced as if it were burned in my stove.***

Life energy in animals and plants

Oxidizing carbon in a living animal also releases CO_2 and H_2O into the atmosphere, but by different chemical means and in several complex, enzyme-driven steps. The energy is released more slowly than in a flame, but it keeps the body warm and is partly converted to mechanical energy that activates the muscles, both voluntary and involuntary.

The biochemical action, in both animals and plants, happens inside the cells. A cell is basically a tiny bag of jelly-like chemical soup that permits — or in some cases drives — fluids, nutrients, and wastes to move in and out of the cell as needed. Cells of animals and plants are similar in chemical composition, including the nucleus containing the genetic material (DNA), whether in yeast or grass or worms or lobsters or finches or giraffes or my neighbor across the fence.

Plant cells operate in exactly the same way as animal cells, using carbon as the fuel for their metabolic needs. But inside the plant cells are also little packets of chlorophyll, the green stuff so prominent in vegetated landscapes. During daylight, the chlorophyll, by wondrous molecular maneuvers, captures the energy in sunlight. The cell uses that energy to rearrange the atoms in CO_2 from the air and H_2O from the soil into one or more of the carbohydrates described above — wood fiber or some form of food. This is the exact reverse of the oxidation of fuel in animals, and after consuming the carbon and hydrogen, plants release the leftover oxygen. This is why ***we owe a debt of gratitude to plants for every morsel of food we eat and for every breath we draw.***

The additive nature of burning

Because my firewood pile grows smaller as winter progresses and my gas tank gradually empties as I drive along, the perception of burning is that it is a process of subtraction or reduction. But that perception deceives us: burning is additive, in that the result of burning is actually larger and heavier than the fuel that was burned. We just don't notice it because that result is in the form of gas and vapor.

This is most easily shown with simple math. The atomic weights (relative weights of the atoms) of the three players are:

Carbon: 12
Hydrogen: 1
Oxygen   16

When I drive along burning gasoline, the molecular weight (combined atomic weight) of the invisible CO_2 gas that flies off into the air is

$$C\ (12) + O_2\ (16 + 16) = 44$$

For the hydrogen part of the gasoline, the molecular weight of the water that also vanishes into thin air is

$$H_2\ (1 + 1) + O\ (16) = 18$$

A gallon of gasoline weighs about 6.3 pounds, 87% (5.5 pounds) of which is carbon. If all that carbon is oxidized into CO_2 gas, then the weight of the CO_2 gas is greater than the weight of the gasoline by a factor of $44/12 = 3.67$. This oversimplified math says each 6.3-pound gallon of gasoline converts to, among other things, around 20 pounds of CO_2!

It's also worth noting that **converting those 5.5 pounds of carbon to CO_2 consumed about 14.5 pounds of oxygen.**

My car gets about 25 mpg. If I drive for one hour at 50 mph, the car's engine inhales and exhales several thousand times a minute and burns 2 gallons of gasoline, producing about 40 pounds of CO_2 — which, at sea level and 70°F, would fill a cube about 7.5 feet on each edge — while I inhale and exhale about 15 times a minute and, burning yesterday's toast and corn flakes, exhale about 1 cubic foot of CO_2. And the same is true for the car ahead of me and each of its occupants, and the cars behind me, and each in the stream of cars passing me, and those on the next road over, and so all over the mechanized world. Each of those drivers can look in their rearview mirrors and see no trace of the change they have made in the atmosphere. But it is huge.

CO_2 in the atmosphere

In 1958 Charles Keeling took his newly invented device for very accurate measurements of CO_2 in the atmosphere, and he went to a forest campground to test it. For a few weeks he took CO_2 measurements every 4 hours, day and night. He found that every morning the level of CO_2 in the air began to decrease, and in the evening it began to rise again. As the daily rotation of the earth carried the local forest eastward into daylight, the green stuff — trees, shrubs, grass — inhaled CO_2 until nightfall. In other words, ***the forest inhales and exhales once each day.***

Keeling then took his device to the observatory on the top of Mauna Loa, the highest point in Hawaii. Mauna Loa is still a somewhat active volcano, but the prevailing winds are always west to east, and the observatory is well upwind from the volcano. This should be some of the cleanest air on the planet, passing over a thousand miles of ocean since last leaving any population centers, traffic, or power plants.

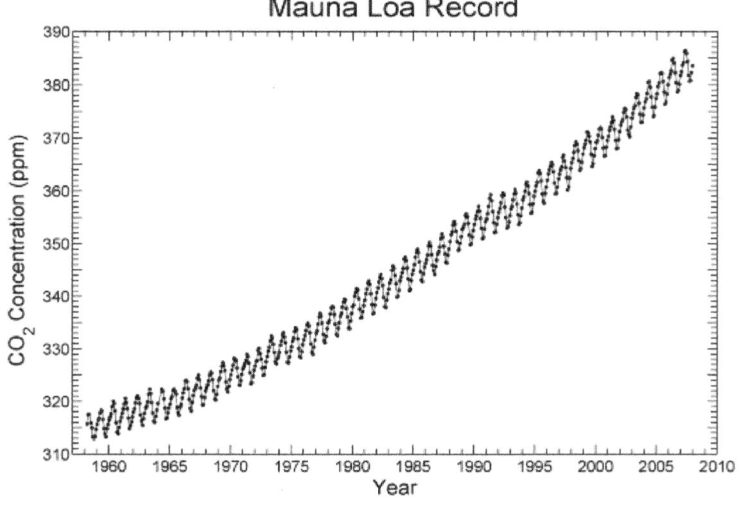

Keeling's curve

Keeling's measurements showed regular annual variation: the CO_2 level rose to a peak in April, then steadily declined until October, when it again began to rise.

We might expect that result. As the earth, in its annual orbit around the sun, tilts its northern half toward and then away from the sun to produce the seasons, the greening of leaves carries through from spring until autumn, April to October, reducing the CO_2 in the atmosphere; and the CO_2 begins to increase as plants enter winter dormancy. In other words, ***the entire North Temperate Zone inhales and exhales once each year.***

But the most strikingly visible result of Keeling's measurements is that, from 1958 to 2010, the average and peak CO_2 levels for each year have steadily risen, at a gradually accelerating rate, from 315 parts per million (ppm) in 1958 to 390 ppm in 2010. This overall rise is what currently is known as "Keeling's curve."

Though it has taken two centuries for our coal- and oil-burning industrial era to add 100 ppm of CO_2 to the atmosphere, the level is now rising at an ever-increasing rate.[9] This is the kind of "positive" feedback that is visible in many other aspects of our current civilization.

In May 2013, the average level of atmospheric CO_2 as measured at the Mauna Loa station passed 400 ppm, the highest since the Pliocene Era about 3 million years ago.[8]

The greenhouse effect inevitably follows from the rise of CO_2 in the atmosphere. The most familiar example of that effect — other than a greenhouse itself, whose heat-trapping ability allows us to grow plants in winter — is a car parked in the sunshine with the windows closed. Sunlight enters through the glass and is converted to heat, and the heat is trapped inside by the glass. The heat that builds up in that car can be deadly.

The human contribution

I often hear it asked, "Why do so many people still dispute the science of global warming and the part that humans play in causing it?" Indeed, five of the U.S. congressmen elected in 2010 scoffed at the whole idea.[10]

My answer is that **they are not disputing the science, but rather dismissing it.** After all, much of it is the same science, the countless global measurements of air and water temperature, humidity, barometric pressure, satellite images, and similar computer models that give us the massive weather charts and predictions that flood our television screens daily — tomorrow's snow, a hurricane forming in the Gulf, tornado watches across southern Kansas, and probable frost tonight in my back yard.

But it's so counterintuitive. Because of age and health I have given up driving, so standing here and looking smugly around, I don't see any sign that I'm warming the planet. But then I think of all of the things in my kitchen and how far they were transported — using fossil fuel — from the picking to my table: tropical fruits, coffee and tea and chocolate, grains, all from thousands of miles away. On average, food in the U.S. travels approximately 1,300 to 2,000 miles from harvest to the consumer's table; in the United Kingdom, total food travel is about 18 billion miles per year.[11]

The list of ecological degradations that include my small personal share goes on and on. I can deny or rationalize it, but I have no place to hide.

For those who may still doubt the human role, we can start from the other end. Pretend for a moment that all of the unprecedented increase in atmospheric CO_2 over the past two centuries came from wildfires, volcanoes, and other mysterious natural

The fact that a significant part of the CO_2 increase is due to human activities is further shown by isotope studies that can identify CO_2 from burning fossil fuels and distinguish it from that from wildfires and other non-human sources.

sources. We know that, regardless of where CO_2 comes from, it is removed from the atmosphere mainly by chlorophyll in trees, grass, bushes, or the algae in the top few feet of the oceans.

We know for certain that human activity has decimated the earth's green stuff — 96 million hectares, about the size of California, of clearcut rainforests just in Brazil; two thirds that much in Indonesia and similar amounts throughout the tropics;[11] good farmland paved over throughout the world because that's where the people concentrate. And 500 miles off California's coast is a vast, slow-moving whirlpool, the North Pacific Gyre, rotated by ocean currents. It is also known as the North Pacific Garbage Patch because the rotation has concentrated floating trash — mostly plastic and styrofoam, as well as tires, bottles and cans, bits of wood, oil and chemical sludge — in an area six times the size of the United Kingdom. The trash on the surface reduces the light that reaches the algae below, so they also are reduced and can't remove as much of the CO_2 that has dissolved in (and acidified) the water. This huge garbage accumulation is the result of human activities at their finest.

By removing plants and shading algae, we humans have reduced the capacity of our world to absorb CO_2 and convert it to oxygen.

Sunlight is the fuel and chlorophyll (the green stuff) is the carburetor driving the engine of all life. We humans have greatly depleted the chlorophyll. Now the temperature gauge is in the red, smoke is coming from under the hood, and unfamiliar noises are emanating from the engine. What does a prudent motorist do? Thus far, our response is to continue increasing our speed. We desperately need to apply mental and societal brakes rather than the accelerator.

Homeostasis

Systems and feedback

By "system" we mean an arrangement of things functionally related to each other, or interacting with each other.[12] Certain basic principles are involved in the identity of any system, be it a cell, a toad, a person, a school, a Boy Scout Troop, a nation, or a solar system.

One of those principles is that systems are constantly changing, whether growing, wearing out, running down, reproducing, or disintegrating. A system cannot continue on and on without change, though the changes may be exceedingly slow.

Another general principle in all systems is that the changes must be restrained. Mechanisms must be built into the system to prevent any change from exceeding the adaptive power of the system. Otherwise, the system will ultimately be destroyed by its own action.

Yet another principle is that every system consists of subsystems, all the way down to subatomic particles, and every system is part of a larger system, all the way up to the entire universe. This idea of congruency helps us understand the relationships between and among systems.

Physiologist Walter Cannon,[13] impressed by the resilience and effectiveness of the regulatory controls in the human body (regulating temperature, pulse, hunger, breathing, and much more), coined the term "homeostasis" for this process of allowing for change but keeping it within the limits of survivability.

Homeostatic mechanisms are what modern engineers call feedback loops or closed-loop control systems.[14] A thermostat, any sort of guidance or steering system (human or mechanical), or the population of rats allowed to breed freely in a confined space are all examples of the feedback control principle. Each change results in other changes which tend to counteract the first change, thus limiting the amount of deviation that can take place. In that way, the temperature stays within a survivable range, the vehicle moves along its slightly zig-zag course, and the population of rats levels off at a point where the death rate equals the birth rate.

Another kind of feedback process has the opposite effect. (It's often called "positive" feedback, but its effect usually isn't positive.) One change sets up a series of changes that in turn reinforce the original change, producing an ever-increasing escalation. This sort of feedback effect, if unchecked, will eventually destroy the system.

The mechanism of homeostasis is perfectly illustrated in the Eco-jar. For example, oxygen consumers (creatures such as snails, shrimp, or crabs) produce CO_2, and CO_2 consumers (plants) produce O_2. To survive, the two must achieve a balance. If either the CO_2 or the O_2 in the jar becomes depleted, a new balance must be struck, and that usually means something must die.

An example is the self-escalating process of combustion, heating a bit of wood enough to cause a flame, which produces enough heat to ignite the next larger stick, which in turn ignites more wood. This process continues until the house is burned up, thus halting the feedback.

Feedback in human affairs

A primary characteristic of our present social and economic system is our attitude toward growth, progress, and expansion. ***We measure the***

health of our economy not by its size, but by its increase in size in a given period. A company that fails to grow in a year has by definition had a bad year. The objective of corporations now is not only the increase of profit, but also the increase in size and corporate power.

We have developed a vocabulary that in itself is a persuasive force toward continuous acceleration of change. *Progress, expansion, growth, development, under-developed, primitive, aboriginal, backward,* and *stagnant* are all words that perpetuate the emotional commitment to doing more of what we are already doing, and they have a powerful feedback effect that is largely hidden.

The result of these value assignments is a self-escalating "positive" feedback process. As was pointed out above, this sort of feedback effect, if unchecked, eventually must destroy its system. A pebble starting down a loose mountainside sets in motion a larger stone, which in turn starts more stones rolling. The resulting avalanche is finally terminated by the counteracting feedback of reaching the valley floor. That stops the process, but its effects must then be absorbed by the homeostatic processes of a wider system.

If one system collapses, it is swallowed up by the next larger system, thus triggering some homeostatic adjustment by that system. This means that, in varying degrees, everything is functionally connected to everything else. You can visualize this kind of interdependence much more easily in the Eco-jar than in a larger setting.

Another force for escalation is the self-fulfilling prophecy. Having once predicted a certain trend for the future, we then gear our present actions to that expectation, thus helping to ensure the prediction will in fact come true.

Similarly, we could project into the future our present growth-based approach to living until some limiting factor or process — perhaps catastrophic — halts and reverses the changes. If for some combination of reasons humans should become extinct, the homeostatic adjustments made by the remaining life forms in the larger Earth system would result in a different ecosystem, and the dance would continue.

Homeostasis as a creative principle

Thus far I've presented homeostasis in its traditional sense as a protector or stabilizer of systems that already exist. The Eco-jar is host to innumerable homeostatic feedback loops among the various inhabitants and the other parts of their overall system. But homeostasis goes far beyond this, and it can illuminate some larger truths.

No two creatures are identical, which means every creature is born with a difference. Some differences render a creature less compatible with its surroundings, and some give it special survival advantages. It is these interactions not only among creatures, but also among internal biological subsystems, that make evolution a selective process.

Suppose I create six new entities, each with its own unique set of properties. I create those properties by giving an instruction card to each and an item to all but one of them. Here are the properties of my six new entities:

1. This person has no item and behaves as follows:

Look around the room. If you see a flashlight on, take one step toward it and turn your thumbs down. If the light is off, point your thumbs up.

2. This person has a flag and behaves as follows:

Look around the room. If you see someone with thumbs pointing down, take one step toward them and lower your flag. If their thumbs are pointing up, raise your flag.

3. This person has a flashlight and behaves as follows:

Look around the room. If you see someone holding up a flag, take a step toward them and turn your flashlight on. If their flag is down, turn your light off.

4. This person has a pair of scissors and behaves as follows:

Look around the room. If you see someone with a flag raised, open and close your scissors. If their flag is down, stop moving the scissors.

5. This person has a card with a green end and a red end and behaves as follows:

Look around the room. If you see someone with a flashlight turned off, hold your card by the green end. If their light is turned on, hold your card by the red end.

6. This person has a walking stick and behaves as follows:

Look around the room. If you see someone with a flag raised, tap your stick on the floor. If their flag is lowered, stop tapping and close your eyes.

Now imagine these six entities scattered randomly among a large crowd, say at a cocktail party. Entities 1, 2, and 3 will eventually end up close together and functioning as a closed-loop feedback system, each alternating their actions indefinitely. Entities 4 and 5 will respond to the actions, respectively, of entities 2 and 3, and will continue doing so, but they will not become part of nor have any effect on the closed-loop feedback system among the first three. Entity 6 will respond once to entity 2 and then become disconnected from the system.

The result illustrates the creationism of evolution, and casts it in a whole new light.

Symptoms

Economics as a branch of psychology

The nature and value of money

We treat money as something with intrinsic value, like a nourishing breakfast or a warm blanket. But you can't eat a plate of coins, or wrap yourself in $20 bills, so where exactly is the value of money?

Imagine this conversation, a very long time ago:

Bo: *"I didn't do well at hunting today. I'll give you 3 arrows for that extra antelope you killed."*

Ab: *"Done. I know you make good arrows, and I have plenty of meat."*

A week later:

Bo: *"Hunting has been good, but many of my arrows were lost or broken and I haven't had time to make more. Can you spare 3 of yours? I'll give you meat."*

Ab: *"Sure. Here, take the arrows. But I don't need meat right now."*

Bo: *"All right. I'll lend you these 3 sea shells, one for each of your arrows, to remind you of my promise to give you meat when you need it."*

Ab: *"That's a handy idea. I'll put them on a string and hang them in my cave."*

Our economy began with barter, the most direct agreement on the relative utility of things. But because the timing of needs doesn't always match the timing of availability, like ripening of fruit or success in the hunt, some token of mutual trust between the parties is used in place of the actual goods.

When I say I "have money," I really mean that I have accepted the loan of tokens, or reminders, of my trust that Bo, or some anonymous person in the future — someone who is also a member of what we might call the same Mutual Trust Society — will give me something with actual utility, like meat or arrows or a shirt or a tank of gasoline, in exchange for those tokens.

The "value" of something, contrary to our usual understanding, is only and exactly what someone is willing to pay for it, and that may change over time. Thus the entire money system is only as stable as the solidarity of the mutual trust throughout the system. But it seems we have lost our understanding of this societal relationship with money.

It once seemed strange to me that, during the Great Depression, people were suddenly hungry, homeless, and unemployed, and people previously rich were in despair and jumping from tall buildings. *The previous prosperity was gone, yet the basis of that prosperity was still in place.* The soil in the fields was still there, still warmed by the sun and moistened by the rain, and the metal ore was still in the mines. The machinery of prosperity lay idle while its skilled operators stood in line at soup kitchens.

What had gone missing — the key ingredient — was the mutual trust. It took some wrenching changes, and finally a war — overwhelmingly seen by the public as a defensive and therefore a morally valid war — so that everybody rallied around and all worked together, gradually restoring the mutual trust.

The same loss of trust was apparent during the economic collapse of 2008 and thereafter. A constant and universal inflation of expectations, combined with the finest entrepreneurial cleverness and competition, finally over-inflated the balloon of expectations in most of the industrialized countries worldwide, so nobody could tell whom or what to trust. It was the latest triumph of globalization.

Economics, psychology, and complexity

The basic, hand-to-hand money system has developed into an increasingly complex "science" of economics, including again the assumption that money, or capital (accumulated money), has some intrinsic utility rather than being a promise, like the original sea shells, to provide something of value at some future time.

To confuse this issue further, we now have technology by which money has been made entirely imaginary. I know of an aging miner who, like Rip van Winkle, not long ago went to the bank in his small town with a quart jar full of gold dust and nuggets, and wanted to sell it, just as he had done decades ago. He was directed to a local trader in precious metals who, in exchange for the gold, gave him paper tokens of trust in the agreed-upon value of raw gold by weight.

Mutual trust is the platform, and hope, fear, and greed are the legs on which the presumptions, or mythology, of economics stand.[15]

We may consider a retail grocery or hardware store, an auction, and a casino as three points along a spectrum of economic psychology from simple to complex

In the **retail store,** the pieces of merchandise are visible, with information about each, and prices are fixed and indicated on each item: take it or leave it. The customer's psychological exercise is to weigh his perceived need (or want) against the cost, measured against his resources, and to decide whether he "can afford" the exchange. The merchant's psychology is more complicated, and is of course directed to maximizing his profit. Included will be advertising, tricks of merchandising such as product placement and visibility, eye-catching displays to encourage impulse buying, competition with other merchants, and numerous other decisions.

The ***auction*** has different psychodynamics. The auction company selects the type of objects to be sold — from expensive artwork, as with

Christie's or Sotheby's, to collections of various kinds and price levels (*i.e.*, the range of what someone may be willing to pay), to estate sales, to common miscellany near the lower end of the price range, such as yard and garage sales. The auction client will seriously and carefully survey in advance the objects to be sold, select what he would like to buy and what he would be willing pay, and then will stick to those decisions once the bidding starts.

The **casino** involves a third set of psychologies. The games are designed so that the odds favor the house. Some games are purely a matter of chance, at least from the player's standpoint. Some, such as poker and blackjack, add a limited element of skill to the presumably random dealing of cards — skill in remembering what cards are visible and what cards have been played, and speed in calculating how those factors affect the ever-changing odds. Thus the psychology of the player may include hope; denial that the odds are against him, perhaps coupled with greed if and when he wins some; willingness to spend some money for the entertainment, and anticipation; and perhaps desperation.

The stock market is a sort of hybrid between an auction and a casino. The trust waxes and wanes, with hope, greed, and fear — even unto desperation — never far below the surface. Yet the public still appears to believe that the stock market, like money, is intrinsically real. "The Market was nervous yesterday after Friday's unemployment numbers." "The Market has struggled to cross the 10.000 psychological barrier." "The Market roared toward record territory, heartened by congressional recognition of ..." It is an object, almost, of worship. Yet a person could "make" a million dollars one day and lose it all the next day, and not even be aware of it. It would have no effect at all on their grocery bill or even their mortgage payment.

Other traditional economic institutions, such as banks, loan companies, investment firms, insurance companies, and pawn shops, all have their various psychologies, which are not hard to see if you imagine yourself on one side of the transaction and then on the other.

Reflexivity

Most "disciplines" or subjects of specialized knowledge have their specialized practitioners, people who presumably understand the topic better than ordinary people and who believe — and enforce — whatever is the current mythology of the specialty. In this respect economics is no different, but the "science" of economics is much different from the physical sciences. It is not subject to proof or refutation by any kind of experiment, but rather rests on assumptions, ideology, and interpretations of past events.[16]

George Soros, one of the richest men in the world, is an economist who has beautifully explained the difference between the economic and physical sciences with the concept of "reflexivity."[17] In physical science, the reality is not changed by the scientist's thinking. Kepler and Galileo and Newton did not change the orbits of the planets by observing them, much less by thinking about them: the movement of information was a one-way street. But if a person is a participant in the thing they think about, their thinking inevitably affects and becomes a part of the subject. Thinking about economic processes and values is reflexive, and becomes part of a new reality. In a reflexive process, information flows in both directions, so that the "reality" is always in flux.

Seduced by the "Great Frontier"

Depletion of natural resources in the Old World around the Mediterranean and the Middle East had brought hardship and poverty, with wide disparity between the royalty and upper economic classes and the average farmer or tradesman, for several centuries. With the discovery by European navigators of the resource-rich lands of the New World — North and South America and the Caribbean islands — a whole new frontier appeared, a seemingly limitless supply of both new and familiar resources.

The economy of the early American colonies was almost entirely agrarian, with simple trades that supplemented farming.

Economist Herman Daly[18] points out that most economists throughout the 1800s showed an awareness that the earth and its resources are finite, and therefore would eventually place limits on growth of population and of per capita use of resources. But as the North American colonies grew, the unmeasured vastness of the continent seemed to promise unlimited land, rivers, forests, space, and opportunity. If someone felt crowded in the colonial towns, the advice was, "Go West, young man!"

Though the new society was almost entirely agrarian, the industrial revolution was beginning to accelerate. Craft tools — hand tools used by blacksmiths and furniture makers and tinsmiths and weavers and gunsmiths and the like — were augmented by early tools such as water-driven looms and spinning wheels and flour mills, and then by standardized parts interchangeable from one firearm to another, and the era of tools-to-make-tools accelerated.

Adam Smith, in his 1776 book "The Wealth of Nations," popularized the view that in economic matters, if each individual is motivated by his own individual interests, a sort of "invisible hand" would cause the overall result to bring the greatest good to the public interest. This became the leading force behind the idea of the "free market," or *laissez-faire* economics — that is, without any government interference.

This mindset has become firmly entrenched in the psychology of modern economists. Adam Smith's "invisible hand" gave rise to such slogans as "A rising tide lifts all boats," along with the "trickle-down theory" that benefits to the wealthy are benefits to all society. This further transformed into the idea that economic success will come to anybody who really tries, and that the poor are just potential millionaires who have somehow missed the train. The value placed by economists on individual private wealth has become, many would say, the primary and dominant value in our culture, almost equivalent to a religion. As a result, the "public good" is viewed with suspicion as an attempt at transferring wealth downward, raising the specter of a socialist or communist dictatorship.

As industrial technology in mining, communication, transportation and more picked up momentum, economists shifted their attachment increasingly to the policy of continued growth. They developed, as Federal Reserve chairman Alan Greenspan once said of investors in the stock market, an "irrational exuberance." They assumed that by maximizing both supply and demand, production and consumption would feed each other in an ever-rising spiral, and this would naturally mean increasing the common good, raising standards of living, and increasing life satisfaction for consumers indefinitely.

Stepping into Alice's Wonderland

The mindset of economists today, reflected in the behavior of the entire industrial system, is that the highest good for the public is maximum throughput in the entire system. The assumption is that this is the key to our national virtue and strength as a superpower, and that it will automatically improve the lives of everyone.

This world view is supported by economists' illusion that ***scarcity of physical resources is no longer a threat,*** because we now have, or can find, a technological fix for any problem if necessary. We can find previously unknown deposits, create a new substance to replace the depleted one, invent new ways of doing without it, or we can decide to deal with it later.

In 1976, the Journal of Economic Literature[19] contained this statement: ***"Man has probably always worried about his environment because he was once totally dependent on it."*** This would be amusing if it were not so extremely dangerous.

The dominant goal of economics now — as constantly called for in the public media and most elements of society — is continued growth of both production and consumption, *measuring the common good by the rate at which industrial busyness and consumers' buying habits grow ever larger.*

Economists, having devotedly bought into this new dogma, and being the anointed "scientists" of their discipline, are de facto the priests of this religion, analyzing and advocating the most effective ways in which to advance their goals. Thus, being the primary advisers to the corporate world, they set the guidelines for maximizing growth and profit for the whole global corporate structure — and, by extension, for the mindset of the entire culture.

The ecological costs of this policy, because they are difficult to quantify in economic terms, are considered "externalities," collateral damage, to be dealt with by somebody else — specifically, by Adam Smith's "invisible hand" — so that, to an amazing degree, the hazards of escalating ecological scarcity have disappeared from the view of most economists, hence from much of the public as well. Thus the worldwide engine of devastation continues to pick up speed.

The steady state

Walter Cannon[13] coined the term "homeostasis" to describe the dynamic equilibrium maintained in the body. Economist Herman Daly has long preached principles similar to those of Cannon as applied to economics, and also to the industrialized human uses of the Earth's physical resources. He calls his principle "steady state economics." [19]

Daly's "Steady state economics"

Daly goes to great effort and trouble to define what he means by the steady state, and also what it is not. He stresses, for example, that many modern economists have tried to include in the definition of steady state situations in which the rate of growth and the rate of change remain about the same. Daly prefers to call this "proportional growth" rather than a "steady state of growth." In his introduction, Daly summarizes his general position:

"Economic theory ... has assumed that environmental sources and sinks are infinite relative to the scale of the economic subsystem. ... [But] the economic subsystem has grown to the point that its physical demands on the ecosystem are far from trivial. We have moved from a relatively empty world to a relatively full world [in terms] of human beings. ... Beyond some point — perhaps soon for developed countries, and ultimately for all countries — economic growth is both physically and economically unsustainable, as well as morally undesirable."

Some critics see the arguments for steady state economics as a policy for despair and willing acceptance of the problems that beset us now as a result of constant economic growth and the concomitant excesses of consumption and production. Daly rejects that interpretation of the meaning of steady state economics.

Daly also quotes John Stuart Mill in defining what he means by steady-state economy:

"It is scarcely necessary to remark that a stationary condition of capital and population implies no stationary state of human improvement. There would be as much scope as ever for all kinds of mental culture, and social progress; as much room for improving the Art of Living and much more likelihood of its being improved, when minds cease to be engrossed by the art of getting on."

Economics of the Eco-jar

Probably the easiest path to a clear understanding of the "steady state" concept is to reconsider the Eco-jar. If we accept that the jar is an authentic model of the Earth's Life System — in that it is a closed system except for input of sunlight and out-radiation of heat — then, after a period of adjustment, the situational economics in the Eco-jar are as close as you can get to a steady state.

From the moment the Eco-jar is sealed, its ingredients are fixed and cannot increase nor decrease. They can only change from one form or location to another, balancing and reorganizing in response to their conditions and interactions. Likewise with the contents of the Earth's system: take a little from here and move it over there, but the total amount has not changed. Burn it or reshape it or dissolve it in acid, but the total amount remains the same, changed only in form or location.

The changes or movements in the Eco-jar are motivated by the unique properties of each element in the jar, and by solar energy. But there's one primary difference between the Eco-jar and Earth's Life System: the presence and influences of humanity.

Human input to the Life System begins with the assignment of meaning and value to various elements or combinations of elements.[20] We then alter their forms and/or locations in accordance with those meanings and values.

Some changes, in the Eco-jar and in the Earth system, may make elements no longer useable or available. In both systems, once a plant or creature dies, it decomposes into its components. Although those components are available to the rest of the system, that plant or creature will not live again in its original form, nor therefore will it contribute its life functions (respiration, consumption, excretion) to the system. In the Earth system, the lead in gasoline or any substances flushed into the ocean are, for current and practical purposes, irretrievably lost,[21] even though they are still present within the system.

The remarkable evolution of the corporation

"I see in the near future a crisis approaching that unnerves me and causes me to tremble for the safety of my country. . . . Corporations have been enthroned, an era of corruption will follow, and the money power of the country will endeavor to prolong its reign by working upon the prejudices of the people until the wealth is aggregated in a few hands and the Republic is destroyed."
— Abraham Lincoln

Corporations, or some equivalent entities, have existed for millennia, probably since hunting and gathering were replaced by the beginning of non-nomadic agriculture, in the Fertile Crescent of the Tigris and Euphrates rivers 16,000 years ago, and in Egypt some 5,000 years ago.[3] This change led to social stratification, economic hierarchy, and political and bureaucratic structures. Corporations were a major instrument in Europe's exploration, trade, and colonization during the 15th and 16th centuries.

Most of the original American colonies were corporations chartered by the British or other European colonial governments. For more than a century before the American Revolution, the colonies of Connecticut and Rhode Island were each governed by a "Governor and Company," incorporated by charter from the English crown.[22]

In those days, a corporation was a group of individuals authorized by law to act as a unit, as an arm of a government for a limited purpose. It had certain specific attributes:

1. ***It was identified as a "person"*** — a fictitious person, to be sure, intangible, but with some of the characteristics of an actual person.

2. ***It had perpetuity;*** that is, its members might come and go, die, or be replaced, but the corporation continued.

3. ***It had a name,*** separable from the names of its members, in which it could sue or be sued.

4. ***It had the right to hold property*** as its own, which was not the property of any of its members.

5. ***This property was not liable*** for the obligations of its members, nor was their private property subject to obligations of the corporation.

6. ***The corporation had a well-defined constitution,*** or rules set by its charter. It dealt with minor internal matters through By-Laws that were set by the members but could not conflict either with the charter or with the laws of the land.

7. It was a long-established British principle that ***a corporation could not create another corporation.***

The word "corporation" doesn't appear in the Constitution. The powers of the Federal Government were enumerated, and all other powers were to be exercised by the States. Thus the chartering of a corporation became a function of each state. The states, with the experience of having themselves been corporations, were familiar with the concept and with the traditional attributes of a corporation, and they chartered such corporations as seemed appropriate.

The era of petroleum

In the 1860s, with drilling and refinement of crude oil, there began a rapid transition in home lighting from whale oil to the much less expensive kerosene, and in 1868 Standard Oil (Pittsburgh, Pennsylvania) was chartered under the ownership of John D. Rockefeller, who hoped to cash in on this booming new business.

In 1886, in one of a series of Federal Court decisions, Santa Clara County vs. Southern Pacific Railroad sharply expanded the protections of the "personhood" of corporations under the Fourteenth Amendment, which was originally intended to protect the rights of recently freed slaves:[23]

"No State shall make or enforce any law which shall abridge the privileges or immunities of citizens of the United States, nor shall any State deprive any citizen of the United States of life, liberty, or property without due process of law, nor deny to any person within its jurisdiction the equal protection of the laws."

It would be a long time before similar personhood rights were guaranteed to women, African Americans, or men without property.

Also in 1886, Karl Benz's patent on an automobile made gasoline — a previously nearly useless byproduct of kerosene production — into a whole new financial frontier. In the years following, numerous "Standard Oil" companies were chartered in other states under Rockefeller's ownership. At first the name of each corporation clearly indicated its home base. Some of these, however, bought up smaller oil companies, sometimes invoking the Standard name and other times retaining the name of the acquired company or giving it an entirely new name, so that keeping track of their identities and market areas required a complicated map and catalog. Rockefeller organized these numerous oil companies into a Standard Oil Trust, the better to consolidate his power, and other competing companies jockeyed for market opportunities as well.

In 1911, under provisions of the Sherman Anti-Trust Act passed earlier by Congress, the Supreme Court ordered the break-up of the Standard Oil Trust — and the stock prices rose sharply as the sum of the parts became worth substantially more than the Trust itself.

The next phase of corporate culture

With this background, and the rapid proliferation of corporations thus established, let us fast-forward a half-century to the words of John Kenneth Galbraith:[24]

Few subjects of earnest inquiry have been more unproductive than study of the modern large corporation. The reasons are clear. A vivid image of what **should** *exist acts as a surrogate for reality. Pursuit of the image then prevents pursuit of reality.*

With the corporation's designation as a legal person, the people, by court actions and in practice, have created a metaphor and made it to act as if it were animated material. Yet the "legal person" of the metaphor differs from a flesh-and-blood person:[25]

The most obvious property of the corporate person is the absence of a body. It has, therefore, no forbears, no heirs, and is immortal. It has neither youth nor adulthood, but only chronological age and size. It is spared all corporeal (as opposed to corporate) fears of injury or death. It may be robbed but not raped, confiscated but not killed. Without eyes, it can know neither beauty nor ugliness, nor can it read. Without ears, it knows neither sound nor silence. Lacking genitalia, it is immune to the joys and blandishments of sex. Lacking lungs and liver, it cannot strangle on its own effluents.

... Strictly speaking, a corporation can "act" only in a narrowly restricted sense. It may "own," "arrange for" (legally) or "conspire" (illegally). It does not build, it arranges for building to be done by its flesh-and-blood agents. It probably should be considered able to buy, sell and sue. It cannot commit crimes of violence, and imprisonment is not among its spectrum of hazards. It can be guilty of crimes, however, and a variety of criminal acts may be committed in its name.

We can carry this pretense so far as to speak of the "parent corporation," but its offspring are not children, but rather subsidiaries, often non-profit foundations or other showpieces serving mainly public relations purposes. The term "subsidiary" is merely a statement of ownership, and the roles of owner and owned can be reversed by a simple stock transfer.

The "personhood" of the corporation thus described could apply as well to a robot, which can perform some of the functions of a human, but those functions must be derived from the imagination and mechanical cleverness of the robot's creator. By its nature the corporation, like the robot, can have no conscience, no generosity, and no morals, and is responsible only to its owners — the stockholders — and to the laws of the land.

The corporation can, however, easily obscure the stark and cold rationality by which it operates by summoning the metaphor of a "good corporate citizen." A vice-president joins the Rotary Club, or chairs a committee supporting a United Way fund-raiser, and the public takes this as evidence of corporate civic values and concern.

Of some 200,000 corporations in the U.S., 80% have been called "entrepreneurial."[25] These corporations are owned by their stockholders or investors, and they resist any constraints on their activities by government or regulatory agencies. They do most of the ordinary work done in the country.

The other 40,000 are more or less "mature" corporations that operate in a very different manner. The larger ones, especially, are heavily dependent on government support of various kinds — financial support of research and development, both in independent laboratories and through universities; support especially for weaponry, presumably for U.S. defense, but actually potentially for arming many other countries, even if they are adversaries. Their corporate identities can easily be changed at their discretion. Such a corporation can change its legal nationality by establishing subsidiaries in another country or by buying a foreign company, and, if it will gain tax or market or public relations advantage, it can further obscure its identity by changing the name of the new company. ***Thus the*** *power of such a mature corporation supersedes all of the legal structures as well as the boundaries of any country.*

The thrust of change in corporations, especially in the past three decades, has been in effect to blanket the earth with their growing influence. There is now more cooperation than competition among global corporations. They form an almost unbroken web of control over the world's trade, money flow, and resources, while becoming so integrated into cultures and public awareness as to become unnoticed, like background noise or the force of gravity — a pervasive element of the cultures of all the industrialized countries.

Corporate predation

The next corporate mutation, beginning in the 1990s, is described most vividly, passionately, and comprehensively by Vandana Shiva.[26] Shiva was one of India's leading nuclear physicists, but she gave up that career to become an activist and advocate for the peasant farmers, the subsistence agriculture, the traditions, the rich heritage of biodiversity and natural resources, and the vast local knowledge among indigenous societies about the biological resources in the Third World countries. She stresses the scant appreciation and respect for women and their contributions to society, including being custodians of much of the knowledge about the innumerable species of edible, medicinal, and other usable types of plants and animals.

She points out that the tropical areas of the Earth have by far the greatest biodiversity and the greatest biomass, especially in rain forests, and also the greatest abundance of the natural resources needed to maintain the lifestyles of the industrialized nations in the temperate zones. The tropics also have the lowest degree of industrialization, and the lowest standard of living by Western standards. This combination of traits makes the Third World peoples a tempting target for exploitation by the industrialized nations, especially as their own resources become ever more depleted.

The "free trade" movement is the latest step in development of mature corporations as the ultimate predators. As Shiva points out, part of the genius of this gradual expansion of power was the use of the word "free," which further obscures what is happening.

The free trade movement probably should be thought of as starting with the Breton Woods negotiations after WWII, which initially dealt with liberalizing economic burdens on the vanquished countries.[27] The International Monetary Fund (IMF) and the World Bank were early products of the Breton Woods deliberations. Later (1948) came the Geneva Agreement on Tariffs and Trade (GATT), part of which was the concept of "restructuring." This meant that any loan or grant of foreign aid to a poor country generally came with the requirement to restructure the country's political and social institutions by privatizing government programs — selling or otherwise transferring to private companies such programs as health care, schools, water systems, and sanitation. The new owners could then change or abolish or consolidate the services, and charge individual user fees for schools and health care. The money collected went into the coffers of the corporations involved, often leaving the already struggling governments to deal with the financial wreckage.

This practice, with various elaborations, has continued to expand and spread under the World Trade Organization (WTO), which replaced GATT in 1995; the North American Free Trade Agreement (NAFTA), Central American Free Trade Agreement (CAFTA), and other continuing expansions of trade arrangements. Because these are international agreements, no one country can go against their provisions without facing sanctions from the compliance arm of the WTO.

Aristotle said more than 2,000 years ago that because of scarcity and other ecological problems, "slavery may be necessary for civilized life."[28] One must wonder whether, in our relations with Third World peoples in general, GATT is the modern counterpart of that slavery.

Wade Davis[29] points out that about 6,000 languages are currently spoken on the Earth, and each language represents a unique worldview. A large number of them are spoken but not written, so their history, their tribal identity, and their knowledge about their part of the natural world have been preserved — in some cases for many centuries — in stories, songs, rituals and customs. We in the developed countries have been able to use our languages to accumulate and deploy technical knowledge and power as a weapon against the indigenous peoples of the world. The very word "indigenous" acknowledges that these people were the original inhabitants, hence the original "owners" of the land and resources, before our crews arrived.

Corporate power over life forms

When Europeans first colonized the non-European world, they felt it was their duty to "discover and conquer," to "subdue, occupy, and possess." It seems the Western powers are still driven by the colonizing impulse to discover, conquer, and possess. And in addition to controlling every society and culture, our corporations now seek to conquer the interior spaces, the "genetic codes" of life forms from microbes to plants and animals, including humans. In 1996 a U.S.-based corporation patented the breast cancer gene in women in order to gain a monopoly on diagnostics and testing, and this started a trend. Up until 1996, certain cell lines of the Hagahai of Papua New Guinea and the Guaymi of Panama were patented by the U.S. commerce secretary.

Biopiracy is the Columbian "discovery" 500 years after Columbus. Patents are still the means to protect this piracy of the wealth of non-Western peoples as a right of Western powers.[26]

It is enlightening for me as a Westerner to hear first-person voices for Third World experiences and viewpoints. Vandana Shiva is an especially articulate example, probably the best because of its international scope. A few others I recommend are Winona LaDuke,[30] Debra Harry,[31] and Ian Mackenzie.[32]

Historically, I am born of that minority of people historically driven to "discover and conquer, subdue and possess every society, every culture." Our history books don't put it that way, and none of us now alive remembers the historical "colonizing days." Unfortunately, these Third World voices show us that it still goes on now under a different guise — "intellectual property rights."

One of the ramifications of the free trade agreements is the laws and patents protecting intellectual property rights (IPRs). The notable rationale for this is to prevent electronic piracy of such things as music or movies or books or electronic technology. However, the most hideous of these developments is extending IPRs to the patenting of life forms.

Selective breeding by traditional methods, generation by generation, is simply an acceleration of the natural selection process that has always driven evolution. But transferring pieces of DNA in the laboratory, from the nucleus of a single cell from one species into a cell from another species, is totally different. It produces an organism that has not been naturally tested and selected by exposure to a natural environment. There is no way to test in advance whether this new plant or animal may have disastrous effects on existing populations, or on itself.

In Canada it is no longer possible for a farmer to raise canola that he can prove is not a genetically modified organism (GMO). That's because Monsanto Corporation's canola seed, modified to be resistant to Monsanto's plant killer (Roundup), has run loose in the environment — meaning neighbors' canola fields — by wind-blown pollen or by seeds being carried by birds or on boots or tires.

Here's the headline:[33]

Monsanto wins lawsuit over pollen. *A judge on Thursday ordered a Canadian farmer to pay thousands of dollars to the biotechnology giant Monsanto because the company's genetically engineered canola plants were found growing on his field, apparently after pollen from modified plants had blown onto his property from nearby farms.*

Vandana Shiva speaks at length about the inroads of Monsanto and other corporations into the agriculture of India.[34] Farmers were pressured and persuaded to buy the seeds of the "Green Revolution," Monsanto's seeds that had again been genetically engineered to make them resistant to Monsanto's own Roundup weed killer. Also required were pesticides and artificial fertilizers in order to get the full production from the new seeds.

The costs of this program required the farmers to borrow money — a totally new experience for them. Because the plants produced sterile seeds, they were forbidden their age-old custom of planting seed from the previous crop. So buying new seed each year from Monsanto added further to their debt. Their despair and humiliation drove hundreds of them to suicide.

Vandana Shiva, in a 1997 address in Boulder, Colorado, quoted a Sandoz representative as saying:

We can't stop to have debates with the public on patents on life and the ethics of it. We can't stop to do an environmental impact assessment of the release of genetically engineered organisms. Because by the turn of the century there will only be five of us controlling health and agriculture and energy, and we have to be one of those five. **If we stop to worry about ethics and ecology and other irrelevant considerations, we will lose out in the competition.**

In Canada, Atlantic salmon have been artificially endowed with genes from a sea perch, which makes them grow to harvest size in 18 months rather than the usual four years. These altered salmon are farmed in pens off the Pacific coast, and already uncounted numbers have escaped from the pens and joined the normal runs of several Pacific salmon species that migrate up Canadian and U.S. rivers to spawn. Will they interbreed? Will all of the species become sterile and salmon become just a historical curiosity? Will some glorious 200-pound monster emerge and create a vast new fishery? Stay tuned for the results over the next 40 years, for we have no way to answer these questions in advance.

Culture or conspiracy?

I made the point in Chapter II that it's not evil people that do these things. I'm sure many, perhaps most, members of corporate business teams go home after work and spend a peaceful family evening without any hostility, much less violence. In the morning, it's back to the office to do the corporate work with pride and satisfaction, and with no thought of the ecological consequences. We have been taught, and the culture has reinforced, that this is good and necessary work.

However, it is difficult to avoid thoughts of conspiracy and evil when people appear to be bragging about just that. For example, here's David Rockefeller, speaking at the June, 1991 Bilderberger meeting in Baden, Germany:

We are grateful to the Washington Post, the New York Times, Time Magazine and other great publications whose directors have attended our meetings and respected their promises of discretion for almost forty years. It would have been impossible for us to develop our plan for the world if we had been subjected to the lights of publicity during those years. But now the world is more sophisticated and prepared to march towards a world government. ***The supranational*** *sovereignty of an intellectual elite and world bankers is surely preferable to the national auto-determination practiced in past centuries.*

It is clear that the distinction between cultural traits and conspiracy is a very faint line: one man's political strategy is another man's conspiracy.

How to respond?

We saw in the previous chapter that economists, being official custodians of the knowledge about commerce, money, and wealth, are the ones who draw the flow charts and road maps for corporations. Partially blinded by the delusion of infinite resources, and guided by Adam Smith's "invisible hand," economists have greatly assisted the evolving nature of corporations.

Corporations have gradually morphed from being a limited arm of government with a specified function, to independent entities with lives and purposes of their own, to eluding any constraints by any established political institutions (cities, states, nations), to finally co-opting the governments themselves and using them to protect and feed the corporate goal: extracting maximum profit from any possible resource on the planet.

This is a swarm of parasites devouring their host. If it were happening in the Eco-jar, the green stuff would be tuning brown and the water becoming cloudy, signs of gradual death of the Life System.

In "Wisdom of the Tools,"[25] Bill Merrill says:

Society does not dare tamper with the large corporations, except peripherally. This is not because of possible retribution by furious managers. It is simply that the work they are doing is deemed utterly necessary, and that there is no alternative mode available for getting it done.

I now believe what Merrill failed to grasp: that a large part of the work these corporations do, which we deem to be utterly essential, is work that **should not be done at all.**

The commons

Natural self-interest and resource consumption

Long ago, when the world was young and people were few, everything belonged to everybody — the air and water, the sunshine and rain, plants and animals. The people moved as they wished, and they learned how to live in the places where they were. No one thought of "owning" any of it except personal or family items like clothing and tools and weapons that could be carried with them as they moved from place to place. Everything needed for life was provided by the ecosystem.

As people gathered together in nomadic tribes or settled villages, and domesticated animals, it was still understood that all life came from the land and that everyone needed the things it provided. Farther on in history, as populations increased and political institutions developed, even if the king controlled the landscape and collected taxes from the serfs, the land was still known to be the source of all goods. The people took their animals to the pasture to graze, and there was still enough for everyone.

Garrett Hardin[35] describes the natural progression in this way:

Picture a pasture open to all. It is to be expected that each herdsman will try to keep as many cattle as possible on the commons. Such an arrangement may work reasonably well for centuries because tribal wars, disease, and poaching keep the numbers of both man and beast well below the carrying capacity of the land. Finally, however, comes ... the day when the long-desired social stability becomes a reality. At this point the inherent logic of the commons generates tragedy.

*As an **apparently rational** being, each herdsman seeks to maximize his gain. ... He asks, "What is the utility to me of adding one more animal to my herd?"*

*Since the herdsman receives all the proceeds from sale of the animal, the positive utility is nearly +1. The negative component is a function of the additional overgrazing created by one more animal. But the effects of overgrazing are shared by all the herdsmen, so the negative utility for any decision-making herdsman is only a fraction of –1. Thus the **supposedly rational** herdsman concludes that his only sensible course is to add another animal to his herd. And another; and another. But this is the conclusion reached by each and every herdsman sharing the commons. Therein is the tragedy. Each man is locked into a system that compels him to increase his herd without limit — in a world that is limited. **Ruin is the destination toward which all men rush,** each pursuing his own best interest in a society that believes in the freedom of the commons. Freedom in a commons brings ruin to all.*

The obvious way to avoid this Universal Ruin is to place political or institutional constraints — either coercion or incentives — to curb free use of the commons for private gain. But as the commons become more and more divided up and controlled by private owners, the economic value of the remaining commons increases so its pursuit is even further intensified.

It is said that when you have a hammer in your hand, everything looks like a nail. Corporate economists, with their new perception of unlimited global resources, feel freed from the constraints of ecological scarcity. **What once was a list of dwindling resources now looks like a global shopping list.**

Corporate raids on the commons

Consider water, one of the most basic examples of the commons. The Earth's water is constantly on the move. Ever-expanding rivers carry rainwater from higher to lower elevations; lakes receive water from rivers, then pass it on downhill to the ocean. Water also is constantly raised from the ocean, lakes, and rivers by evaporation and carried by the wind back to the headwaters of the streams, to fall again as rain. Some water follows the same cycle over again, but some seeps into the ground and recharges the aquifers, which can extend for hundreds of miles underground before resurfacing.

The water pays no attention to the dotted lines on a map, and the would-be users of water simply look for whatever access to water they can find. A city such as Las Vegas, having been founded in an arid area, has run out of ample water and is feeling the effects of impending shortage; and many of the world's most densely populated areas have suffered extensive drought.

Even as I write this, the remaining commons are being captured and divided up by private entities. An example was published in Newsweek's cover story on Oct. 18, 2010, "Liquid Asset: Big Business and the Race to Control the World's Water." Sitka, Alaska, whose population of 10,000 is scattered over 50,000 square miles, has one of the world's purest lakes, fed by snow melt and glaciers. "Every year, as countries around the world struggle to meet the water needs of their citizens, *6.2 billion gallons of Sitka's reserves go unused.* That could soon change."

A U.S. company, True Alaska Bottling, has purchased rights to transfer 3 billion gallons of water per year from Sitka's Blue Lake. Another company, S_2C Global, will ship the water in tankers to their processing plant in Mumbai. Sitka hopes thus to reap a $90 million industry. The water companies claim that market forces, guided by Adam Smith's "invisible hand," will best determine the price of the water. Water thus becomes a global commodity, for sale to the highest bidders.

Of course, ***those who most desperately need the water are least likely to be the highest bidders.***

James Olson, an attorney who specializes in water rights, points out that "Water has been a public resource under public domain for over 2,000 years. Ceding it to private entities feels both morally wrong and dangerous."[36]

Water merchants are a diverse lot. Multinational water giants Suez and Veolia deliver water to some 260 million taps around the world. Former oil wildcatter T. Boone Pickens wants to sell the water under his Texas ranch to Dallas and other thirsty cities.[37] In the 1990s the World Bank required Bolivia (among many other countries) to privatize their water system in exchange for economic aid. Bolivia's water prices doubled, citizens rioted, and in 2000 Bolivia's government took the water system back from Bechtel, who had been operating it.

In addition, the water commons is used for disposal. Rivers, lakes, underground aquifers, and oceans are increasingly fouled by sewage, toxic chemicals, fertilizer and pesticide runoff; silt from topsoil erosion, mining, and dredging; chemical and oil spills; and wastes deliberately injected into wells to force out valuable oil ("fracking").

Another kind of commons has very different implications. Instead of constantly falling from the sky, purified by distillation, the non-renewable minerals — and fossil fuels — have been deposited underground over eons of geologic time, and nature has no process of replenishment short of new geologic processes. Thus ***simply to estimate the remaining world reserves*** — of tin, manganese, tungsten, copper, aluminum, and all other metals, as well as fossil fuels — ***implies the same mindset as that of a suicide bomber:***[38]

 1. Intentionally bringing about the end of the world.

  a. The bomber by blowing it up — himself included.

    b. The industrialist by using it up — along with his sources of food and
      air.

2. *Justification:*
    a. The bomber to kill as many infidels as possible, and then to go to a
      better place that's equipped with 72 virgins.
    b. The industrialist to use it up, convinced that a technological fix can be
      found when it becomes imperative.

As the commons in the industrialized world become more and more completely enclosed and under the control of private corporations, the quest for resources reaches farther into the parts of the world inhabited by indigenous or subsistence populations, who most often reap the environmental and resource costs but not the benefits of corporate incursion. In northern Canada and the Arctic, controversy over mines or proposed mining projects and oil and natural gas drilling is widespread among the First Nations people, who understandably have no confidence in oil companies' ability to clean up a spill under Arctic ice.[39]

It is increasingly obvious that we are experiencing on a global scale what Garrett Hardin's essay predicted: "Freedom in the commons brings ruin to all."

Exponential change

At the mention of "exponential curve" many people's eyes will glaze over. Even when depicted on a graph as an ascending curve, exponential change is counterintuitive and severely stretches the imagination. But to see the scope of our present predicament as a society, and even as a species, we must comprehend viscerally the power of exponential change, the effect of repeated doubling of any quantity.

Linear change is produced by repeatedly adding or subtracting a number. Exponential change is produced by repeatedly multiplying or dividing by a number.

Examples of exponential change

To comprehend the reality of exponential change, let's look at several examples.

Horseshoe nails

A farmer went to buy a horse from his neighbor. After the horse was selected, the buyer balked at the mention of the price.

"All right," said the seller, "We've been neighbors a long time, so let's do it this way. The horse is shod, and there are six nails in each shoe. Just give me one penny for the first nail, two cents for the second nail, four cents for the third nail, and so on, doubling each time, all the way to the last nail. I'll accept that amount, and we can still be friends."

Each got out his calculator and went to work.

The left front foot of the horse came out to be worth $0.63. "This could be embarrassing," thought the buyer.

Adding the figures for the right front foot, the price was up to $40.95. "This might work," muttered the buyer uneasily to himself.

When the tally included the right hind foot, their calculators each said $2,632.43.

The buyer stared at the final number, then at his neighbor, and said aloud, "Weird." After he silently left for home, the seller turned back to his calculator and shook his head in amazement at the final price of the horse: $671,089.31.

Such is the power of exponential change — repeated doubling, or repeated multiplying by even a small positive number, as in compound interest.

Global population

I arrived on the Earth in 1923. At that time, after many thousands of years and perhaps 65,000 generations of humans, the population of Earth had reached about 1.5 billion. ***Before I turned 70, the global population had doubled twice,*** and at this writing it is almost 7 billion — in less than one lifetime!

Moore's Law

"Moore's law" says the number of semiconductors that can be put onto a 1-square-inch silicon chip doubles every two years. This has continued to be true for more than 46 years, or 23 doublings. (Remember the horse? The result of 23 doublings is more than 330,000 times the original number.) This has led to a torrent of digital technologies and devices that penetrate every aspect of our culture. It has transformed our communication, transportation, education, production and consumption, entertainment, weapons systems, international relations, art, and cultures around the world. And these devices become obsolete almost before they reach the markets.

Forty years after the publication of his law, Moore estimated that two more doublings would be the limit.[40] "It can't continue forever. The nature of exponentials is that you push them out and eventually disaster happens."

Wheat on a chessboard

The king offered a rich reward to his court wizard for a service. The wizard replied, "Sire, I don't ask much. Just give me one grain of wheat on the first square of the chessboard, two on the second square, four on the third, and so on, doubling the grains on each square. This is all I ask."

A week later the royal treasurer brought the final tally: after the 63rd doubling (next to last), the chessboard would contain more wheat than the entire Earth's current annual crop, and the final pile would occupy 10 cubic miles.

It is clear that "the earth could not sustain 64 doublings of even a grain of wheat."[18]

Lily pads on the pond

We are very suddenly on totally uncharted ground. Consider a man who constructed a pond of about 1 acre in size, in which he planted his favorite strain of Japanese koi, a carp-like fish. All went well until one day he saw a strange water lily in one end of the pond. After extensive research, he learned that this was an invasive strain of lily that would double in size every day, and when it eventually had covered the entire pond it would block the light, suffocating the entire fish population.

After a few days he noticed that the patch of lily pads had enlarged noticeably at the end of the pond. Clearly he would soon have to take action to remove the lilies, but he was busy with other things. A few days later he saw that the pond was ¾ covered by lilies. So how long does he now have in which to clear out the lily pads to save his fish?

Right. He has only today.

Real exponential change

Exponential increases occurring in our world today include the following:

- per capita consumption
- pollution of rivers, oceans, and aquifers
- deforestation
- loss of topsoil from erosion and agriculture processes
- species extinctions
- genetic manipulation
- increase in atmospheric greenhouse gases (CO_2 and methane)

The list goes on and on, and the speed and extent of these changes is unprecedented in human history. The last two generations of humans have never experienced the fabric and texture of culture which the previous thousands of generations — including mine — took for granted as "the real world."

Petroleum

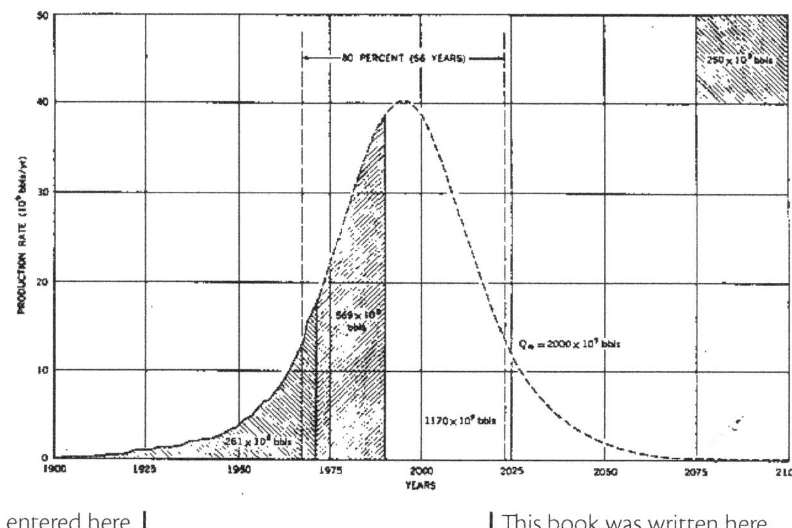

I entered here. | | This book was written here.

One of the many things that have changed exponentially over the years is our extraction and consumption of crude oil and its components. This graph shows the course of that process. *It also shows that I have personally lived through the consump-*

tion of more than 80% of all the oil that can ever be produced, from its very beginning to the last dregs available.

"Wow!" you might say. "Who's the crazy doomsayer who came up with this?"

I'll let him tell you who he is, as he told the Subcommittee on the Environment, of the Committee on Interior and Insular Affairs, House of Representatives, in hearings held June 6, 1974. His credentials are too extensive to record in their entirety, but he provides the essentials:

"My name is M. King Hubbert. I am a Research Geophysicist with the U.S. Geological Survey. My scientific education was received during the 1920s from the University of Chicago, from which I have received the degrees B.S., M.S., and Ph.D. jointly in geology and physics with a minor in mathematics. One half of my professional career, beginning in 1926, has been in both operations and research with respect to the exploration and production of petroleum. The second half has been divided about equally between university teaching in geology, geophysics, and mineral and energy resources. ... In the petroleum industry my work included geological and pioneer seismic explorations in Texas, New Mexico, and Oklahoma during 1926-1928 for the Amerada Petroleum Corporation, and in petroleum exploration and production research during l943-1963 for Shell Oil Company. I helped to organize and staff a major research laboratory for petroleum exploration and production."

After further description of his study and teaching about different patterns of growth, Hubbert goes on with the business of the hearings:

"Of particular pertinence to the present hearings on the rate of industrial growth has been a continuing study, begun in 1926, of mineral and energy resources and their significance in the evolution of the world's present technological civilization.

"Two terms applicable to an evolving system are of fundamental importance. These are steady (or stationary) state and transient state. The growth phenomena with which we are at present concerned are almost exclusively of the tran-

sient kind. Three types of transient growth are illustrated in Figure 1 [below]. This figure is drawn with a time base extending from the year 1800 to beyond 2100, during which some quantity is assumed to grow in one of the three modes shown. The first of these growth modes, shown by Curve I, is uniform exponential growth. In this curve the magnitude of the growing quantity is assumed to double every 20 years.

"A second type of growth is that shown in Curve II. Here the growing quantity increases exponentially for a while during its initial stage, after which the growth rate starts to slow down until it finally levels off to some fixed maximum quantity. After this the growth rate becomes zero, and the quantity attains a steady state. Examples of this kind of growth are afforded by biological populations and by the development of water power in a given region.

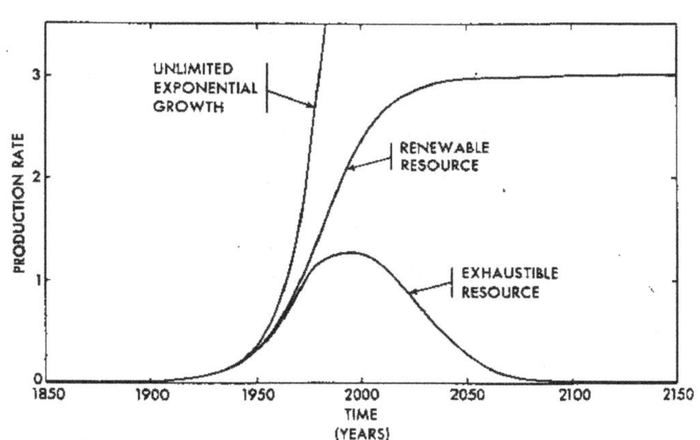

The population of any biologic species, if initially stationary, will respond to changed conditions in a manner indicated by Curve II, or conversely by its negative analog. That is, the population in response to a disturbance will either increase exponentially and then level off to a stable maximum, or else decrease negative-exponentially and finally stabilize at a lower level, or perish. The development of water power in a given region behaves in a similar manner. The curve of installed capacity finally levels off and stabilizes at a maximum compatible with the potential water power afforded by the streams of the region.

*"A third type of transient growth is that represented by Curve III. Here the quantity grows exponentially for a while. Then the growth rate diminishes until the quantity reaches one or more maxima, and then undergoes a negative-exponential decline back to zero. **This is the type of growth curve that must be***

followed in the exploitation of any exhaustible resource such as coal or oil, or deposits of metallic ores."

Extrapolating Hubbert's pronouncements

Over the past million years or so, despite a gradual increase in density and geographical spread of the human population, energy use per capita — in the form of food derived from solar energy by photosynthesis — changed very little. The ecological system of the human species can only be regarded as a slowly changing steady state. Although the pace quickened about 8,000 to 10,000 years ago with the domestication of plants and animals, a rapidly changing transient state of evolution was not possible until mining of coal as a continuous enterprise was begun near Newcastle in northeast England about nine centuries ago. This was followed as recently as 1857 in Romania and 1859 in Pennsylvania by the pumping of oil from wells.

"Whenever history repeats itself, the price goes up."[3] When past civilizations have destroyed themselves by consuming or destroying their natural resources, such as the Sumerians, the Maya, and later the Easter Islanders, people elsewhere in the world were not affected, and in fact never even heard of the events. However, the next time it happens, the collapse will be global and everyone will be affected.

Let us now celebrate with gratitude and humility the Whole Life System, ancient and indivisible, and the "I" that briefly dwells therein.

Ecological scarcity

Soil

The basic ingredients for all food are topsoil, rain, seeds, and sunshine. (This is also true for meat-eaters. Remember the words of Isaiah: "All flesh is grass.") Since the beginning of agriculture some 16,000 years ago, a farm has been a net producer of energy. The domesticated plants and animals assembled the above ingredients to produce the food for the humans, and food for the draft animals that supplemented the human muscle work, and almost all the other necessities of life. This was all true at least until after 1900, when automobiles and then farm tractors made their appearance. Many farmers — including both sets of my grandparents — never did shift from horses to motorized machines, and they actually lived off the land itself.

Over the next several decades, however, almost all farms shifted to tractors. Now most farms are net energy consumers, meaning that the calories of petroleum energy put into the farm are more than the calories produced. Oil not only goes into gasoline and diesel fuel. Large amounts also go into the manufacture of poisons against weeds and insects, and the conversion of nitrogen into a fertilizer form accessible to plants. Thus, although the petroleum-intensive method permits much larger operations, it is — in the terminology of physicists or engineers — much less efficient than were the farms of my grandparents.

Wes Jackson is a plant geneticist, president of the Land Institute in Kansas, and one of the godfathers — along with farmer and author Wendell Berry — of the sustainable agriculture movement. Jackson believes the most important piece of our vast ecological heritage is topsoil, and the greatest ecological threat to the global food supply, and indeed to human civilization, is the steady loss of topsoil due to erosion.[41] He points out that this has been an inevitable result of agriculture — tilling the soil — ever since it began in the uplands of Iran thousands of years ago. In agricultural areas the rivers are always muddy with the particles of soil washed into them, "headed for the ocean, never to return." And natural processes can take 1,000 years to replace a foot of topsoil.

Soil is also carried away by the wind. In the 1930s, the effects of severe drought plus the moldboard plow ripping up deep-rooted native prairie grasses resulted in the Great Plains becoming the dust bowl, and topsoil from the middle of our nation was blown west "to Washington and even to ships at sea." Each year when farmers in China begin plowing, dust particles appear at the observatory on Mauna Loa, thousands of miles to the east.[11] Wes Jackson says, "If you're a hunter-gatherer, it's pretty hard to cause serious ecological trouble, but once you start tilling the soil, you have become a troublemaker. Agriculture depends on depletion of the Earth's capital stock." It has been said that the greatest discovery of the 20th century was the complexity of the organisms in the soil, and that "only those who know the most about it can understand how much is not known about it."

In addition to erosion, industrial-scale agriculture is functionally depleting the topsoil by the application of pesticides, herbicides, and chemical fertilizers, destroying most of those crucial restorative organisms so the soil becomes little but a container for the chemicals. It will no longer sustain normal growth of crops without continued application of the chemicals. The life of the soil can return, but it may take several years.

Back when our children were children, to expand their experience and education and sense of responsibility, we bought a horse, a gentle old mare. After a couple of rainy days I watched Dandelion plod across the hillside below our yard. Each hoof, as she set it down on the soggy surface, shoved a gob of muddy soil a few inches down toward the river that ran a couple of hundred yards below. I suddenly realized that a hill is a very valuable thing, that mine was undergoing a slow death, sliding imperceptibly down into the river, and that I was responsible for this unintended vandalism. I often walked on that hill myself, and I had put the horse there. A general principle, a mantra, rose in my mind: *When you dig, throw the dirt uphill.*

Much later, on a different hillside, I insisted on this practice with the man I hired to implant a 1,500-gallon water storage tank. He was somewhat annoyed, until he conceded that it made backfilling around the tank much easier — "and more natural."

Water

As we've noted earlier, water is always on the move toward the sea because of differences in elevation. It always is picked up by the air and recycled back to the high places, but the soil particles that have washed down with the water will of course stay wherever they were deposited.

Water not only devises its own pathways on the surface, but it has its own invisible underworld, the aquifers that contain vast amounts of water moving slowly — sometimes at geological time scale — and providing partial support for the ground under our feet. In the 1950s I lived in California's Santa Clara Valley, now known as Silicon Valley but at that time a lush agricultural spread with tomatoes, strawberries, broccoli, "the garlic capitol of the world," prune orchards, and more. Irrigation was intensive, and wells frequently had to be drilled deeper because the water table continually sank lower. Wells near the coast began pumping salt water. In San Jose some cracks appeared in the sidewalks and in the basement of the courthouse, said to be caused by the land over a wide area subsiding a few inches because of depletion of the underground water. Perhaps even worse, more and more aquifers — some extending for hundreds of miles — are becoming contaminated with chemical or radioactive materials that can never be flushed out in our lifetimes.

On the Columbia River on March 10, 1957, the gates at The Dalles Dam closed, and over several hours Celilo Falls gradually disappeared under the rising backwaters. Hear the words of Delbert Frank, Sr., Warm Springs tribal elder:

I used to fish at Celilo Falls before The Dalles Dam was built. We used to be able to fish all year long. We caught lots of different kinds of fish — spring Chinook, summer Chinook, Bluebacks, fall Chinook, steelhead, and Coho. When the fish were coming in good, I could catch one ton of salmon a day. And it didn't take a lot of fancy gear or expensive boats to fish. For the cost of one or two balls of twine, about 6 to 12 dollars, I could make the fishing gear [dip net] necessary for me to catch enough fish to supply my family and many others for a whole year.

Now a series of dams have turned the free-flowing rivers into a series of long, slow-moving pools which disrupt the instinctive downstream movements of the baby salmon at each pause before they enter the torrent through the turbines or into the partially-effective fish ladders.

All over the Pacific Northwest now the lights glow and motors hum day and night with cheap hydropower. But the movement of nitrogen as protein from the vast resources of the northern Pacific up to the clear mountain streams of the Columbia River Basin is a pitiful fraction of what it was when I was a young protein-seeking animal at the headwaters.

We have also profoundly changed the oceans themselves. They are acidified, laced with oil, eutrophic from excessive nutrient runoff, with "dead zones" off the coasts due to pollution or decreased dissolved oxygen.

Globally there are five great Ocean Gyres, very slow-moving whirlpools many hundreds of miles across. The rotation tends to concentrate floating debris toward the center of the gyre. Chapter 6 mentions the North Pacific Gyre, also known as the North Pacific Garbage Patch, some 500 miles off the coast of California. It's a floating mass of debris, some cans and bottles and tires and chemical sludge, but mostly plastic shopping bags or packaging, styrofoam cups, take-out food boxes, and other miscellaneous stuff. This floating dump covers about half a million square miles. It's hard to imagine anyone ever having the means, the finances, and the motivation to try to remove this junk, and it is not biodegradable.

Population

There are too many people. I can tell it just by looking around. But what is my yardstick for this conclusion? Naturally, it is how things seemed when I first was really aware of the world around me — the village streets, the highways (graveled, wide enough to pass but without any stripes), campgrounds, fishing streams. But everything is relative, a matter of comparison. A cattle rancher from my Oregon village of 2,000, on a visit to New York City, remarked on the hotel elevator, "When it gets so there's more people than cows, it's time to go home."

I enter this chapter with trepidation. It is so personal, and its ramifications extend into so many aspects of human values and behaviors: religion, sexuality, gender, race, history, genocide and ethnic cleansing, biology, ecology, medicine, economics, competition, political power and hierarchy, personal and cultural mythology. Always the dilemma of Me as Myself vs. Me as One of Us vs. Them.

Which among us are the too many?

Many people have written about problems related to population — too many people, or too few — as far back as the Garden of Eden: "Go forth, multiply, and replenish the earth."

The physician's dilemma

For me as a doctor, this dilemma has taken on an added dimension. What is expected of the doctor, and what we doctors are taught to expect of ourselves, is to extend everyone's life as long as possible. Though I have never "saved" a life (they are issued only one to a customer), I have nevertheless extended countless lives, by hours or decades, and accepted this as central to my professional duties.

Only later in my life did I begin to realize that I was contributing to a growing problem, not the least by having five children of my own. There is no possibility of estimating how many additional person days or years have been added to the population of the Earth by my efforts.

Herman Daly, addressing the problems implied in matters of population,[42] says the only reasons we have for not destroying Spaceship Earth in an orgy of procreation and consumption are ethical and religious, the obligations of stewardship for God's creation, the extension of brotherhood to future generations, and of some lesser degree of brotherhood to the non-human world.

My own ambivalence — bordering on guilt (as described above) at my professional actions, and partly in response to Daly's statements about brotherhood — is best put in the always difficult and painful context of triage, the old and present populations vs. the far more numerous as yet unborn.

My oldest memory of reading anything relating to this aspect of population was written by William Vogt, an ecologist who observed that India's "most valuable natural resource is the malaria-carrying mosquito, because it keeps the population low enough so that nearly everybody can get enough to eat." And of course malaria strikes most severely on the children. So also do hunger and malnutrition.

I fear that I am on the edge of talking myself into a corner. I could not in good conscience, nor in accord with my basic ethics and sense of morality, support nor appear to condone a policy favoring high mortality among the very young.

Life expectancy

We often hear that life expectancy in earlier times was significantly shorter than it is now. Thomas Hobbes[43] describes life in pre-modern times as "nasty, brutish, and short," and Tennyson[44] refers to "Nature, red in tooth and claw." Industrialized societies have also promoted the assumption that life expectancy and longevity have steadily increased in modern times, presumably as a result of scientific advances, especially in medicine.

Yet, as a lifelong physician, I find these statements difficult to believe. I have no doubt that we live longer in the modern age than we did in the Middle Ages, but I believe we have deluded ourselves about both the degree of and the reasons for increases in our longevity. The intrinsic nature and makeup of the human body cannot have changed so significantly in the last thousand years, and the idea that life expectancy in the Middle Ages was 35 years or less just makes no sense to me.

After some investigation, I've learned that a fallacy exists in the way we normally calculate life expectancy. For a given area and time, life expectancy has virtually always been calculated as the average age at death of all people in that area during that time. Because that average includes infants and children, higher infant mortality — at birth, in the first year of life, and in the first five years of childhood — would drive the average down. Likewise, a reduction in infant mortality would appear to increase life expectancy.

This anomaly conjures up all manner of hypothetical scenarios. For example, if we were to prevent all births for, say, 10 years, we would not only completely eliminate all deaths below 10 years of age, but life expectancy would increase by an astonishing amount. In fact, as a general rule, the fewer children are born, the longer they will live!

To me it makes more sense to ask the age distribution of a population at a given time. Of all the individuals who are living at a given time, how many are infants, how many are young children, how many are young adults, and so on? If a population is suffering high infant mortality, there may be many births and deaths of children in, say, a fifty year period, but they will form only a small percentage of the total number of individuals living at any one time. By this method of assessment, turnover in the population of the young may rise or fall without affecting the apparent ages in the other groups.

What I would like to see is tables that illustrate stable-state age distributions for different societies at different times. Then it would be possible to think of what it might have been like to live in such societies. How many people were, say, over 60 years old? Very few, or a significant number?

H.O. Lancaster provides such a table in his 1990 book, "Expectations of Life,"[45] and I've restated it here. Note that the number in the far column is the number of years a person could expect to live ***after he had reached the age of 21.***

Time Period	Number of Males Observed	Further years of life expected at age 21
1200–1300	7	43.14
1300–1400	9	24.44*
1400–1500	23	48.11
1500–1550	52	50.27
1550–1600	100	47.25
1600–1650	192	42.95
1650–1700	346	41.40
1700–1745	812	43.13
		*black death

From the table it is clear that even in the Middle Ages, if a person could get through childhood and early adulthood, he could expect to live to 64 or so. That also means that ***people older than 64 were living in the society.*** Though the data come from the aristocracy, it is argued that they apply more generally in the society. While the aristocracy are less affected by famine due to their wealth, they are more susceptible to death from serving in the military.

Lancaster's book also says there isn't much evidence of age at death for prehistoric societies, but I have heard of 25- to 30-year life expectancies for early humans. What was likely done is authors have taken modern data on life expectancies, which are on an upward trend (apparently with no end in sight!), and extrapolated backwards. But modern societies have gained increasing average life expectancies mostly by reducing mortality among the young. So my guess is that, if you got in a time machine and traveled back to early populations, you would find lots of older people. Even in the Middle Ages, it simply wasn't that unusual to live to 70 or more.

So don't believe so-called "life expectancies" of 25 years for people in some early society and think they, like many insects, barely had time to procreate before they died. It just wasn't so.

Editor's note

This book was published after Ted Merrill's death on February 26, 2013, and this is the only chapter that was truly incomplete at that time. Although some further exploration of population issues may be buried in his vast collection of handwritten notes, we must settle here for what we have today.

As his eldest son, I've had the pleasure of discussing these issues with him at various times, and I believe an explanation of his thoughts is the best way to close this chapter.

At the beginning of the chapter he expressed his concern at even broaching this thorny subject, and he asked the critical question, "Which among us are the too many?" To my knowledge, he never came to a satisfactory answer for that question. For example, I never heard him say anything for or against Chinese childbearing restrictions, nor did I ever hear him say anything like, "The world would be better off if people didn't have so many kids." Neither of those sounds like an attitude he would have supported.

I first learned of his struggle with this question in 1966 when he returned from a two-month USAID volunteer mission during which he treated Vietnamese civilian patients and assisted local doctors. He was treating people there who were sick with cholera and suffering from malnutrition, and he eventually realized part of the reason he had a constant flow of such patients was that there were too many people for the land to support. He began to question whether curing them and sending them back home was really a positive thing to be doing in that context.

His assessment of the capacity of the land to support the population may or may not have been correct , but this is the way he viewed it at the time. There wasn't enough food for the local people, and their malnourished immune systems couldn't handle the prevalent diseases, so they'd come to him and get some intravenous nutrition and/or be treated for cholera, and then they'd go home and ultimately be another mouth to feed.

For him that situation gradually became a metaphor for his societal role as a physician, and he began — with quite a guilty conscience — to question whether, in an environment overburdened by its local population, it was really a good idea to be curing the very disorders that resulted from that overpopulation. He and I talked about this several times over the years, and the conversations went to issues of Darwinian natural selection and what happens with other creatures when their numbers become excessive for some reason. It's easy to say, "increased predation by wolves controls the elk population," but I don't ever remember him reaching a satisfactory conclusion for the human condition.

Presently, I don't suspect we can all agree on a really satisfactory solution. We humans may have painted ourselves into a corner on this one, and the solu-

tion will probably occur on its own in the form of brutal turf wars and battles for resources when the population, along with other related factors, reaches a tipping point. This is the stuff of epic novels.

Progress ... toward what?

The word "progress" implies moving toward a specific goal or destination. Without an intended destination, progress is no different from aimless wandering or going in circles. If you don't know where you're going, all roads will take you there.

When I begin a trip — unless I'm just out for the ride — I decide where I am headed before I start. If I find that I've taken a wrong turn, I stop. I check the road signs or the map, retrace my steps, find the right road, and move on. When I arrive at my destination, I stop, knowing that I have gone far enough to reach my goal.

The same principles apply in carrying out a project. This was understood even as far back as Aristotle, who said there are four distinct and conceptually essential causes for every rational human activity using material resources: doing what, with what, by what, for what purpose?[46]

We've all heard the common saying, "Necessity is the mother of invention." But the real "necessities" —fire, the wheel, the spear, the knife, the hoe — were invented or stumbled onto millennia ago, and countless conveniences since, and the saying has now been turned on its head. Invention is now the mother, not of necessity, but of manufactured wants. The advertising industry spends hundreds of millions of dollars annually to persuade us that it would be cool and would elevate the quality of our lives to buy, for example, a self-wringing mop: "Call now and get a second mop free! A $50 value for just $19.95!" But they fail to mention that the mop works poorly, or requires extra learning and effort to operate, or breaks the soon after it is activated, and has degraded the quality of life of those who got along just fine with the old-fashioned type of mop and a bucket.

Many philosophers have written on this same issue, for it is by no means a new problem. When the means — more clever gadgets — are treated as ends in themselves, then the real ends — maximum satisfaction in living — can never be achieved.

Changing values and expectations

First, am I happier now than 70 years ago when I was 15? Or am I happier now than my parents were then? Clearly the answer is no. We were generally happy then, and I'm generally happy now. But would I wish to go back to using an outhouse, and a hand pump over the kitchen sink, and heating water on the stove for bath and laundry? No, no, and no.

Why is this so? What has caused my values to change regarding my style of living? I would say that our culture (including me) has suffered a drastic inflation of expectations. We take for granted as necessities things that in a former time we didn't miss because their possibility had never occurred to us. We look around and compare our situation with that of others, and thus we vaguely define what is "normal." Then, as we have learned is laudable or perhaps even obligatory, we try to go beyond what is merely normal — and on and on, determined always to "get ahead."

This behavior in the Eco-jar would obviously mean death for the whole system. If the goal is maximum and ever-growing production and consumption, no destination is possible except catastrophe.

Will our population, our economy, our technology continue to grow in the future? We tend to ask this, and to answer it, vaguely as if it were an automatically guided matter beyond our control. "You can't stop progress." If we continue to see it this way, even when more and more of our "progress" is producing increasing stress on our planet, then it IS beyond our control — which means the human mind, and spirit are more abjectly deficient than I can bear to believe. Our present acts will determine our future, and our present decisions must be made in the context of their future effects as we can best predict them.

We desperately need to revise our cultural idea of progress. By progress we appear to mean to continue on our present course, continually increasing our compulsion to increase production and consumption, continually increasing throughput of the entire industrial system.

Recovery

Back to the Miracle Whip Microcosm

In Chapter 1, "Lessons from the Eco-jar" compared the qualities of the Eco-jar with those of the Earth's Life System; and in Chapter 3, "Perspectives of ecology" added a dimension to those comparisons. Now that we've discussed the symptoms plaguing our Earth, it's time for a closer look inside the Eco-jar.

First, let's review ...

• The Earth, like all the other planets, is a rock circling the sun, but it is the only planet in the universe known to bear on its surface a membrane of life. The Eco-jar, as a tiny slice of that membrane, has boundaries set by the glass rather than by the vacuum of space and the gravitational pull that holds the earth's surface, atmosphere, and life layer in place.

• On the Earth, the Life System is on the outside. In the Eco-jar, it's on the inside.

• The Earth is like a spaceship in that it is a fully self-contained system. It has no access to any outside resources other than the sun's energy and whatever celestial collisions may occur. The Eco-jar, like the spaceship Earth, has no source of resupply for groceries or equipment, and no external waste disposal system. What's there is all you've got.

• The energy by which the Earth's Life System operates comes almost exclusively from the sun's radiation. The only exception is a very small amount of heat from the earth's core, left over from its earliest times. The Eco-jar's energy also comes only from the sun. Because the jar is hitch-hiking a ride on the Earth, its energy intake depends on its surroundings, and it experiences variations in energy input related in part to the Earth's movement: day and night, winter and summer, cloudy and sunny, etc. (See Appendix A for more detail.)

• The Earth's Life System started from scratch on the slowly cooling planet and created itself bit by bit by random mutation, natural selection, and eons of repeated reproduction. The variety and distribution of life forms are determined by their own individual properties and their interactions within the system. The Eco-jar throws together a grab-bag of things already created, living and not, and allows them to self-organize from a mob to a community. Like Earth's system, the variety and distribution of life forms in the jar depend on their properties and interactions. But in the Eco-jar, the self-sorting process takes only weeks or months rather than the original billions of years.

• The Earth is so large in relation to ourselves, our viewpoints, and our movements, that we perceive many separate ecosystems — arctic, equatorial, prairies, rain forests, riparian, desert, estuaries — and we may have to travel long distances from one to another. In the Eco-jar, all the various mini-environments can be seen or imagined. Duckweed leaves float on the surface with their thread-like roots dipping ½ inch down into the water. Hair-like nematodes wriggle and dance just above the mud at the bottom. Surely a whole different environment exists in that muck down there, and in places that are, for example, always in the shadow of a small rock or leaf.

• In the Eco-jar we can easily see that the overall ecosystem is a single interactive and integrated system, that one event directly affects another. The Earth is also a single interactive and integrated system, but we have difficulty seeing this because of its scale of both size and time. Mineral nutrients from deep-sea fumaroles become food for plankton, which die and fall to the sea floor. That sea floor eventually becomes dry land, and the mineral nutrients are lifted up as dust in the equatorial wind, eventually fertilizing plants on another continent and feeding other plankton blooms in other oceans. It boggles the imagination!

- Life is exuberant, but life is also fragile. Within the Earth's Life System, new species of organisms appear — albeit on a relatively long time scale — and existing species become extinct in a relatively short time. In the Eco-jar, extinctions are not only permanent, but also irreplaceable: no new species will appear (unless by mutation of existing, reproducing species). Yet I've had an Eco-jar remain visibly active for 12 years.

Stepping inside the Eco-jar

To gain the fullest benefit from the Eco-jar, we must imagine ourselves completely inside it. Yes, it's small, and the bottom is muddy, and (hopefully) it's got a lot of creepy-crawly things in it. But imagine that we can be inside, and look around at various scales, comparing ourselves with the organisms we see and perhaps putting ourselves in their "shoes."

Imagine you are looking through a microscope. You can see around you the rotifers and paramecia cavorting and grazing, tiny crustaceans, the diatoms with their glass-like shells and green stuff inside, and the various algae. It is quite a varied landscape, perhaps more so than you imagined when you created the Eco-jar.

Remembering that this is a miniature of the whole Earth's system, and setting aside the microscope, you can imagine the plants as trees and bushes and grass, and the other organisms as frogs and eagles and whales, antelope and buffalo, snakes and fish, chimpanzees, and even throngs of humans.

As mentioned earlier, each ecosystem on the Earth — arctic, jungle, deep ocean, savanna, alpine — seems to be a separate system because of the distance between them. But here in the jar, with the scale instantly adjustable up or down by turning a dial in your mind, it is easy to see that the Life System is a single, fully interconnected entity.

Let your imagination focus on the humans in that fully interconnected ecosystem. How do they differ from, and how do they resemble, any of the other creatures? From the bacterium to the whale, from the fish to the elephant, and from the finch to the people, the similarities are striking. Their metabolic chemistry, the range of foods they eat, the principles by which they process it, and the means by which they pass their identities from one generation to the next are nearly identical. The bits are merely arranged in different patterns.

How are we different?

In this entire menagerie, every creature — plant as well as animal — has what can be considered the equivalent of a nervous system: a reaction of a cell to some sort of stimulus from outside the cell. Mammals have generally more elaborate nervous systems than the less highly evolved creatures, primates the most complex, and humans top it off with the final evolutionary addition, the front part of the cerebral cortex and a few uniquely human wiring connections in the brain.

If the "mind," whatever it is, can be found in any specific location, it is here. Its observable activities are generated in the cerebral cortex but integrated with other brain functions, like a symphony orchestra where, for example, the strings and woodwinds stand out but are subtly altered by oboe and flute and grounded on a foundation of bass and percussion instruments.

It is here in the mind that we find the difference between humankind and the nearest likeness, the chimpanzee. The extent of "mind" in the chimp and other animals is still impossible to define clearly, but it is generally agreed and seems obvious that the spectacular difference in culture and behavior between us and the rest stems from language, which makes it possible not only to think in more complex ways but also to exchange our thoughts with others and to pass them on to the generations that follow. For millennia — and in some places today — this was accomplished by stories, song, and ritual. Written language greatly expanded this ability to accumulate knowledge over time, and with successive stages of technology we now are literally swimming in a sea of information.

Among the infinite variety of people in this Eco-jar we call Earth, you'll find engineers, chemists, economists, researchers of all kinds, innovating day and night, competing in the best capitalist tradition.

You'll also find, if you look closely, the Inuit, Polynesians, Athabascans, Penan (what few remain), and some aboriginal peoples in Australia's outback. In fact, nearly 7,000 languages are now spoken on the planet, each representing a unique culture, each with its own mythology, its own world view, its own set of knowledge. But half of those languages are not being taught to children, and most are not written. They are going extinct, the last speaker of the language dying, one about every two weeks.[29] They are dying because their habitats have been invaded and their resources seized by those of us with more powerful tools and weapons.

Those dying cultures have been the most successful on Earth, as evidenced by the fact that they lasted for millennia, carefully maintained in a harmonious way that avoided damaging change to their world.

Human behavior vs. the Eco-jar

In our tour inside the Eco-jar, consider the nature of each living entity, plant or animal, and its interaction with its environment. Then compare those natures and interactions with our human behaviors, keeping in mind that Earth is essentially just a huge Eco-jar.

Competition and cooperation: Which entities in the jar are competing, and which are cooperating? In our human world, competition and cooperation have clear, somewhat opposite meanings. But if competition is an attempt to dominate, can it succeed in the Eco-jar?

Progress and improvement: When you create an Eco-jar, things usually will change rather rapidly at first. You might call that progress because it is evolving toward something. You might also call it improvement, because the ultimate situation will be somehow better (we assume) for whatever inhabitants remain. But in the Eco-jar, progress and improvement are toward a specific goal, the balance among the life processes of all in the jar. Once that balance is reached, progress is no longer productive and certainly will not improve conditions in the jar.

Need, waste, excess, and shortage: In human culture, the concept of waste is one-sided because "one man's trash is another man's treasure." But this becomes a life-and-death matter in the Eco-jar, because waste and need are complimentary. Each entity's waste must fulfill another's need, lest it build up and destroy the balance. Excess and shortage are the two sides of that imbalance, and neither is conducive to life in the Eco-jar. What's required is total recycling.

Hierarchy, predator, and prey: These are concepts we can easily describe on the scale of planet Earth, but within the confines of the Eco-jar there can be no hierarchy, no food chain. Instead, it must be a web, in which everyone is both predator and prey. The big ones eat the little ones, and then the little ones eat the big ones. In my life, I've eaten a lot of little ones, and soon it will be their turn.

Human nature: Consider all this in relation to us humans. Is humanity some sort of special organism? If so, in what ways are we special? Certainly I am special — my dog could not have written this — but my chemistry is practically indistinguishable from his. What makes us "special" is the power of our minds. **Of all Earth's creatures, only humanity has the power, with our minds, to end the whole Life System.**

Community

*"First they laugh at you, then they ignore you,
then they fight you, then you win."*
— *Mahatma Ghandi*[11]

This is not going to be easy.

Ghandi's weapon against the British Empire was the spinning wheel. It was powerful, he said, because it was small. The weapon of the Indian farmers is the seed, also small and powerful, and symbolic of the farmer's place in and dependence on the Earth's Life System.

Integrating work with life

Bill Merrill hauntingly describes our alienation from the Life System:[25]

Picture a man taken from the South African bush, taught "the rhythm of the shovel," and sent down into the mines of Johannesburg, there to work on a meaningless schedule at meaningless tasks, surrounded by swarms of others like himself. From those around him he draws his meager comfort, buries his fears and builds small meanings. For this man, the fears are close to the surface, the unknowable within arm's reach. ... The man with the shovel faces a lifetime torn between nostalgia for a familiar world whose paths his feet know, and the myriad messages he gets from mysterious authorities around him who tell him that his familiar world is dead, valueless and a trap, that salvation lies in the rhythm of the shovel.

But who really owns the shovel? For two million years the Community owned the shovel, and its handle was always warm from the palms of the fathers. Its *rhythm was the rhythm of life, love, and the seasons. The shovel turned up real earth in real mounds for purposes that everyone shared. The labor was not less, but it was life, not labor. The schism separating labor from life decreed the end of man's most successful social form: a small, tribal, non-literate community which had brought him from the primates to his human estate and preserved the species for some 65,000 Homo generations. In the last fifteen of those generations, a driven people from Western Europe has transformed mankind. With irresistible force they have destroyed the community and replaced it with the nation-state, and **substituted progress for life.***

In the small tribal society, the distinctions by which modern heterogeneous society is ordered could hardly apply. Our conventional dichotomies, of work/play or sacred/secular, could have had no meaning.

The contention here is not that life was physically easier, but that it was integrated, undivided. ***All activity took place within a seamless web of legitimacy.***

I believe this touches the very core of our problem, the principal cause of our current suicidal drive over the cliff — the "substitution of progress for life," the separation between life and "having a job." Our first reorientation should be to set a goal, however vague and imprecise at first, of reintegrating work with life, of rediscovering that seamless web of legitimacy.

Self-organizing human communities

In "Harmony,"[11] Prince Charles was struck by the ability of a community, left more or less to itself, to organize itself and improve its quality of life. "Beyond architecture, design, and technology," he says, "perhaps the most important resource in any built environment is the knowledge, relationships, values and perspectives held in communities." He gives two examples, and I have added a third below.

Falmouth

Falmouth is a seaside town in Cornwall, the UK's poorest county. Once a thriving seaport providing thousands of jobs, it gradually deteriorated into unemployment, poverty, drug use and crime. Two health workers tried a new approach, and arranged meetings where residents could describe what they thought had gone wrong, and for the first time in their experience were asked for opinions as to what should be done. A citizens' organization was formed, and a new attitude arose. People planted trees and flowers, painted and upgraded their homes. Over the years unemployment dropped 70% and the crime rate and drug use decreased. This dramatically improved community well-being was achieved, says Prince Charles, "by the strength and power of what had been a latent community capital."

Dharvi

Dharvi, says Prince Charles, is "the largest slum in Asia," 600,000 people crammed into one square mile, poor and homeless people pushed to the outskirts of India's largest city of Mumbai to make way for its great new banking and trade center. Prince Charles, in walking through Dharvi and talking with the people, was impressed with "the vast amount of community capital" by which this discarded and abandoned society organized itself from the bottom up.

Arcata

Arcata is a northern California town in some of the few remaining old-growth redwood forests. In mid-1998, long-time activists Gary Houser and Paul Cienfuegos founded Citizens Concerned About Corporations (CCAC) to launch a local ballot initiative, the "Arcata Advisory Initiative on Democracy and Corporations," named "Measure F" by the county elections office. They believe to this day that it was the first ballot initiative in U.S. history on the subject of dismantling corporate rule.

They received little opposition, most of it a series of inaccurate editorials in the local daily newspaper. They claimed the initiative was "a waste of money" and that the $200 cost to local taxpayers (less than $.02 per voter) would be "better spent filling in a couple of potholes." They also included these tidbits: "Why introduce ordinances that have nothing to do with the way we live our lives or govern our city? ... Why must the city devote any time or resources to a silly little thing like this?"

The initiative sought two things: to run two official Town Hall meetings on the topic, "Can we have democracy when large corporations wield so much power and wealth under law?" and to establish a standing City Council committee that would propose "policies and programs which ensure democratic control over corporations conducting business within the city."

Here again is the power of "community capital." And the parallel between the self-organization in these examples and that in the Eco-jar shouldn't be hard to find.

Measure F won by 58% of the vote, and by early April a new kind of conversation was becoming common in cafes, laundromats, and the line at the post office. In the final days leading up to the first Town Hall meeting, Measure F literally became the most talked-about issue in Arcata.

What does this all mean?

It means a call to revolution. It means — as was argued in Chapter 3 — a total change of values, a new mythology, a new sense of how we fit into the Life System of Spaceship Earth.

It means coming back home, back to the Earth, to the Life System which gave birth to us and which provides our food, water, and breath, hour by hour, whether we acknowledge it or not.

It means each individual connecting with a place, aligning life with the dynamics of a unique region of the earth. Find a place you like, or make your present place likeable. Learn the geography and history and mythology of your bioregion. Find out what you should expect from this particular piece of the earth, and what it expects from you.

It means honoring and respecting your children. Nourish well their bodies, minds, and spirits. The children and the unborn are tomorrow's leaders, and their vision and wisdom are not yet clouded by corrosion from our culture. Teach them both the need and the reason for discipline, gradually internalized as they mature. They are the ones most likely to be able to see the world whole, to make the shift and bring the rest of the culture along.

It means changing our mindset, rapidly and together, without blame and polarization.

It means creating community wherever you are.

The hunter-gatherer standard

Wes Jackson[41] tells how a colleague walked into his office one evening and said, "We need wilderness as the standard by which we measure our agriculture." Then he walked out, and left that concept burning in Jackson's value system.

Obviously we can't all become hunter-gatherers, as there's not enough to hunt or gather anywhere around here to last the community very long. But that ethos, the primitive attitude of dwelling in the whole Life System, and embracing the heart of that concept, should be our ultimate ideal, with labor and living inseparable and unified.

Reflecting, as I so often do, on the Eco-jar, I am ready to echo Wes Jackson's friend and declare, "We need the hunter-gatherer life and mythology as the standard by which we measure our own value system, ethics, and life style."

Electronic devices and children's education

I'm of an earlier age, so I reached maturity before the advent of calculators, computers, and the internet. The industrialized world's dependence on ecommerce is undeniable, but I view the cadre of electronic devices surrounding us primarily as instruments of teaching and learning, and I feel they should be approached with caution.

I personally have found the Eco-jar to be valuable as a teaching as well as a learning device, and most especially as an aid in integrating other things I learn. Consider helping your child set up an Eco-jar and watching the results with them — patiently, for the changes are slow. Encourage them to keep an Eco-jar log or diary. (See Appendix A.)

Teaching devices should enhance a child's mind, not replace it. Children should be comfortably competent at adding, subtracting, multiplying, and dividing on paper and in their minds before they are allowed to use calculators exclusively. In the same spirit, they should be capable of readably hand-writing and expressing ideas with words on paper, with reasonable skill at spelling and grammar, before they are allowed to use a computer for all their writing tasks. Once the mental and physical skills are acquired, the electronic devices can enhance rather than retard the further development of the child's potential.

If you feel comfortable and competent in doing so, consider home-schooling your child. If not, use care in deciding who will do this job, and participate and be involved in a way that supports the efforts of the teacher.

Suggestions for other changes

After all these disturbing discussions, my concerns for the continuing health of Earth's Life System should be fairly obvious. Given those concerns, what would I propose as solutions?

We can make some changes in our individual lives that have ripple effects in our local economy and ecology. Other kinds of changes involve larger societal and political issues. Some of these suggestions may seem untenable or unreasonable, but none are truly impossible. We must summon our collective will to preserve our Life System.

I have tried at various times, in various ways, and with varying degrees of success to make these kinds of changes in my personal life. Any attempts we make along these lines cannot help but have a positive effect.

Personal changes

Grow some of your own food. Even if it's only in a pot on the windowsill, plant something, water and nourish it, harvest it and eat it, and marvel at what happened there. You can find a definite spiritual experience in that process, and your children can learn from it as well.

Eat food that was grown within 100 miles, or as close to that as possible. The Planet Drum Foundation in San Francisco has published a few books[47] dealing with local changes that can enhance urban food sources. Those books promote sustainability and minimal ecological damage on both local and global scales by, for example, replacing grassy parking strips, asphalt and concrete with gardens and orchards. These principles can be adapted in some degree to almost any urban setting.

Slow down. Relax. Walk; ride a bicycle; ride public transportation. Ivan Illich[48] calculated that, historically, when people can travel more than 25 miles per hour, the social fabric begins to break down. Writing some 30 years ago, he noted that the average American drives about 5,000 miles a year, and during that year spends more than 300 hours driving, stalled in traffic, washing and maintaining the car, and working to earn money for license, insurance, parking, and fuel. This all comes to about 15 miles per hour, and you can beat that with a bicycle.

Spend time on the art of living rather than on going. This is part of family and community. The art of living usually includes meaningful interactions with other people.

Practice elegant frugality and simplicity. Distinguish between needs and wants, and consider carefully the best use of your resources. This is the foundation of sustainability and survival, for individuals and for societies.

Remember that you are the owner and the origin of a great amount of community capital.

Pursue the "seamless web of legitimacy"[25] in your life and your choices.

Institutional or political goals

Take part in any political movement toward stability, sustainability, and stewardship, minimizing change and consumption.

Reassert legal control of corporations at every jurisdictional level. See the writings and speeches of Paul Cienfuegos, beginning at www.paulcienfuegos.com.

Oppose aggressive globalization of trade, travel, and finance.

Promote education and incentives toward ***population control worldwide.***

Charge more per unit rather than less for purchasing or using more.

Seek meticulous use of vocabulary, recovering the use of words for information and shadings of meaning rather than as weapons for evasion or persuasion (as in advertising and politics).

Minimize use and meticulously recycle, especially all metals.

Ban genetic manipulation, even for medical purposes.

Ban patents on life forms or genetic patterns.

Require a wide-ranging environmental impact statement for any new area of research in any discipline (including medicine).

Press ahead with development of ***renewable energy.***

Structure your life and situation to ***use only muscle energy*** (human or animal).

Reduce, reuse, and recycle — especially packaging. It's an over-used but important policy.

As a country, ***show a good ecological example*** for the "underdeveloped" nations.

And finally, ***keep in mind the symbolisms of the spinning wheel, the seed, the Miracle Whip Microcosm, and the Clock of the Long Now.***

Averting catastrophe

If enough of these changes can be made to avoid a massive catastrophe, humanity has a chance to achieve, in the words of M.K. Hubbert, "the greatest intellectual and cultural advance in human history."

It finally comes down to this

The Star Trek TV series, some 700 episodes over nearly two decades, inspired a substantial subculture of devoted "Trekkies." Some just loved the science fiction and special effects, but others were drawn to the deep ethical and philosophical issues underlying the story line in most of the episodes. This is crystallized in Starfleet's General Order #1, the "Prime Directive," the basic rule that was to be observed by Star Fleet personnel in intergalactic travels and the search for other sentient life forms:

The Prime Directive

"As the right of each sentient species to live in accordance with its normal cultural evolution is considered sacred, no Star Fleet personnel may interfere with the normal and healthy development of alien life and culture. **Such interference includes introducing superior knowledge, strength, or technology to a world whose society is incapable of handling such advantages wisely.** *Star Fleet personnel may not violate this Prime Directive, even to save their lives and/ or their ship, unless they are acting to right an earlier violation or an accidental contamination of said culture. This directive takes precedence over any and all other considerations, and carries with it the highest moral obligation."*

Ironically, we have done precisely this to ourselves. With our industrial revolution, our enormous technological advances, and the corporate delusion of infinite resources, we have alienated ourselves from the Life System and have "introduced superior knowledge, strength, and technology to a society incapable of handling such advantages wisely" — an unforeseen and disastrous contamination of our own society.

Is it still possible "to right an earlier violation"? In his final words on this final question, M. King Hubbert says that if we can do this in time, we could be on the threshold of **one of the greatest intellectual and cultural advances in human history.**

Appendix

Appendix A:
Creating a Miracle Whip Microcosm

Homeostasis before your eyes

Life, wherever you find it, is a dancing, writhing mix of atoms, molecules, and larger clumps and masses, all interactive and interconnected and interdependent, and all driven by energy from the sun. This is true of the bacteria in the soil, the amoeba and the snail, a pod of Orcas, myself and my significant others, the critters in the Amazon River, the United States Congress, and ice worms on the Arctic ice cap.

Homeostatic ("negative feedback") interactions create relatively stable subsystems within the Earth's great Life System. This can be a staggering concept to wrap your mind around, but you can see it in microcosm, at close range, by creating an Eco-Jar.

Chapter I describes how I created my first "Miracle-Whip Microcosm." Since then I've never been without one or more of these Eco-jars on windowsills in my home. Some fail to thrive and go "dead" (or seem to) in a few weeks or months. One slowly changed for about 6 months, then remained almost exactly the same for 12 years, with a few little swimming creatures smaller than pinheads. Then one day they just weren't there any more.

Here are some tips on how to create an Eco-jar, some lessons I've learned by trial and error, and some ways to avoid mistakes and disappointments I've experienced.

Guidelines for creating an Eco-jar

The location

Find a watery place such as a swamp, stagnant pond, or tidepool. Either fresh or salt water will do, as long as you don't mix them. The water should include at least one kind of plant material and a reasonable amount of standing soil or sand at its bottom. A place where water is standing or moving slowly, rather than rushing quickly or being replaced frequently, is most likely to include plenty of little live creatures in the water, in the soil, and among or inside the plant material.

The plants and animals you have collected are alive, and they need oxygen from the time you dip them up. More than four decades ago, my daughter and I drove several hours home from the Maine coast with a bucket of water containing algae and animals from tide pools. She recalls that her job when we stopped every two hours or so was to spend a few minutes dipping a cup of water from the bucket and trickling it back in from a few inches above the surface to produce bubbles and aerate the water.

The jar

A quart jar is a good size for your first Eco-jar, but a gallon provides a much larger view and potentially a wider range of microsystems. The jar must be made of glass, to allow viewing and to avoid chemical interaction between the water and the sides of the jar. A tight-fitting lid is absolutely essential.

The soil and water

For a quart jar, dip up about half a cup of sand or mud from under the water and pour it into the bottom of your jar. For a gallon jar, up to two cups will be adequate. The muddier the soil, the more likely it is to contain tiny plants, critters, spores, or eggs.

Fill the jar nearly to the top with the water, pouring gently to avoid stirring up the soil too much. Leave about one inch of space between the water and the top of the jar, and set the jar aside to settle. The soil should be allowed to settle in the jar until the water is clear. This may take only a few minutes, but it could take overnight if the soil is muddy.

The plants

In a separate container, collect at least one and preferably two or more kinds of algae or other water plants from the same place. Look for free-floating types, hair-like green strands or clumps, or other types whose stems and leaves are living under the water. Avoid weeds or grass that stand in the water but whose leaves are above the surface.

When the water has pretty much cleared up, carefully add the plant material, enough so it floats free without too much crowding. It should fills much of the jar, but it should also let light pass readily through. This green stuff is a key to the whole system, for it is what captures the sun's energy and provides both oxygen and food for the rest of the society in the jar.

The animals

You may wish to add other creatures you pick up or net from the pond or swamp or tidepool — snails, tiny mussels, shrimp, or aquatic insects. ***Remember to choose only very tiny creatures, and not too many.*** Your little world will be limited, and even an animal as large as a baby crab or small fish would soon use up the jar's resources and die from lack of food and oxygen.

The environment

Leaving an inch or so of air space, put the top on the jar and screw it down tightly. ***It must create an air-tight seal,*** so consider adding a rubber or paraffin ring around the edge.

Set the jar on a windowsill or next to a window. It doesn't have to be in direct sunlight, and in fact, excessive direct sunlight might overheat the jar. But it can tolerate some direct sunlight, and in any case it does need full light from the outdoors for most of every day.

The jar should be allowed to remain motionless most of the time. If you jostle or shake it, the micro-environments of the inhabitants will be rather drastically disturbed, and that is likely to upset the balance that eventually develops among the various elements in the jar. Whenever you move the jar, do so gently and carefully.

Observing the Eco-jar's Life System

Looking through the jar, you can probably see tiny animals swimming around, not only the ones that you added but also many that were clinging to the algae when you scooped it up. With a strong magnifying glass you may see still others. After a while, you may see tiny bubbles forming on the plants and eventually rising to the surface.

People may ask, "Don't you ever have to open the jar to let in fresh air? Don't you have to add food and clean water?" That will be your opportunity to use what you've learned and ask them, "Does the Earth's ecosphere have to have fresh air, food, and clean water added?"

You now have no control over who or what in the jar lives or dies, nor which species of animals or plants will thrive and which will disappear. How long the system will continue depends on many factors, and the only ones you can now control are the placement of the jar, the temperature, and the amount of available light.

If you decide at any time to break the seal and open the jar, to look at it or smell it or take out a sample, you have changed it into a different experiment, and I can't tell you what to expect from that. You're on your own from that point forward!

My own Eco-jars

The jars shown here were set up in June 2000 with stuff from two small roadside ponds, a lake, and a small creek. In the first photo, taken three years later, the appearance has changed some since the beginning. Most of the stem-like fronds of algae have gradually disintegrated, and in Jars 2 and 3 (counting from the left) the water is now very turbid with green one-celled plants. Jar 1 has plenty of live duckweed floating on its surface, but no moving (animal) forms were seen after the first year. Jars 2 and 3 still contained a few little shrimp-like animals less than a quarter inch long.

Jars were set up in June 2000. This photo was taken in March 2003..

Three months after the first photo was taken, a sun-shield screen was left down nearly to the tops of the jars for three weeks. This, along with shade from an apple tree outside the window, apparently decreased the available light too much, and in a few days all the jars began to appear less green, and visible movement in all the jars appeared to have ceased. After the jar was moved into better light, however, the algae in Jar 2 rallied somewhat, and from 1 to 4 shrimp were seen at various times. The last shrimp was seen in Jar 2 in November 2005, more than five years after the jars were established.

In February 2006, at the time of the second photo, Jar 1 appeared totally inert. Jars 2 and 3 contained many barely visible dark colored dots moving about. No other movement has been seen. The plant material appears all but dead — but in bright sunlight, small bubbles of oxygen still form and rise briskly to the surface.

The same jars in February 2006. Two showed signs of life. See chronological notes, page 120.

Getting the most from your Eco-jar(s)

Every Eco-jar is different, and from experience you will have results and learn lessons that I have never encountered. Enjoy, be patient (changes are mostly very slow), and watch always for clues as to how the Earth's Whole Life System operates.

Changes in the Eco-jar are slow and, especially for a child, it's easy to forget what has happened over what period of time. If you place beside each jar a card or notebook and pencil, you can record and date your observations whenever you take a moment to look closely at what's happening inside. Keeping a magnifying glass nearby is also helpful.

Chronological Eco-jar notes

Following are my detailed notes on the three Eco-jars shown in the photos above. The observations began in 2003 when the first photo was taken.

3-22-03

Jar 1. Small jar: Much duckweed, & yellow-green algae on bottom. Water clear. No fauna seen for ~2 yr.

Jar 2. Small jar: two shrimp (seen at same time). Turbid green.

Jar 3. Big jar: One lg. snail, many small snails, at least one shrimp. Clear in top 11" but full of fine hairlike green, 2-3 branches of [hand-drawn: hornwort? like umbrellas (ribs only) strung together]

4/5/03

Jar 2. Marked change in past 2 wks: green color of the water is decreased in intensity, more visibility. No shrimp seen today. Numerous tiny dark critters moving near or on glass in upper 2" (near duckweed mats).

6/30/03

All visible movement gone in all jars. Curtain was down to level of tops of jars, all month of June, -> insufficient light.

5/1/04

Jars appear unchanged (no) except 3 shrimp swimming in middle jar!

7/20/04

Jar 2. One shrimp seen in middle jar.

8/04/04

No shrimp seen today. Plenty of duckweed. Turbid green fluid.

8/18/04

1 shrimp seen. Many tiny oval dark dots moving.

11/1/04

1 shrimp seen. Appears the same: green turbidity throughout, with profuse duckweed.

11/20/04

No shrimp seen.

11/29/04

No shrimp seen.

12/7/04

1 shrimp seen!

1/20/05

4 shrimp!

2/8/05

1 shrimp

2/13/05

4 shrimp.

3/28/05

Jar 1. Yellow-green cotton-like growth (algae) on bottom (~1/4") & suspended from water surface (~1-2"). No movement seen.

Jar 2. Green turbidity. 2 shrimp seen. Numerous tiny, barely visible dots moving. Duckweed abundant.

Jar 3. Tiny dark gray dots & filaments on glass. Light green fluffy algae on bottom, ~1/2 to 1". Numerous tiny moving dots, 3 kinds up to small-pinhead size.

6/23/05

Same as above, plus 1 shrimp.

6/28/05

2 shrimp.

8/11/05

- Neg.

9/5/05

- Neg.

9/23/05

- Neg.

11/5/05

Yes! 1 shrimp.

11/7/05

2 shrimp.

9/3/08

Jar 3. Green filaments throughout: tiny dark dots moving.

12/27/08

Jar 1. Greenish-brown shoots & shreds of algae (or fungus?). No visible movement for few yrs.

Jar 2. Sparse stem-type plants (? elodea). Tiny moving dots, erratic motion. None seen prev. for over 1 yr!

Jar 3. Same as above (9/3/08).

12/27/08

All jars decommissioned.

Appendix B: The NASA/JPL EcoSphere

Several years after my initial Miracle Whip Microcosm experiment, I learned that NASA had, in the 1970s, commissioned the Jet Propulsion Laboratory (JPL) to study the possibility of creating a sealed ecosystem containing plants and animals and operating solely on energy from the sun. Their ultimate goal was to modify this ecosystem to enable interplanetary space travelers to grow at least some of their own food and recycle their wastes.

A higher-tech Eco-jar

In the early 1980s, after extensive research through the south Pacific seeking just the right combination of carefully matched organisms, the technology they worked out was sold to a company called SEBRA, which soon began making and marketing the "Eco-Sphere." This sealed glass sphere was about the size of a softball and contained four half-inch-long red shrimp, a specially selected species of algae, a branch of

The sealed glass SEBRA EcoSphere, ready for shipping.

dead coral for the shrimp to cling to, and a specific assortment of nematodes and bacteria.

These EcoSpheres are now marketed by several companies, and they have been advertised as "the world's only sealed and living ecosystem."

Here is an excerpt from an article in the February 1, 1986 issue of The Oregonian (Portland OR):

The company, which also calls itself SEBRA, began marketing the EcoSphere in October 1983 after buying the technology from the National Aeronautics and Space Administration. The device was developed by scientists at the Jet Propulsion Laboratory of Pasadena, Calif. as part of a long-term research program involving closed ecological life-support systems for space exploration.

The company is willing to talk in general terms. The shrimp are a marine species particularly abundant in South Pacific waters between Chile and Australia. They depend for their captive existence largely on three species of algae. The algae, the solar engines of the little life colony, grow by capturing the sun's energy in its chloroplasts. They consume carbon dioxide from the water and give off oxygen, which is consumed by the shrimp.

Feeding on the algae and bacteria, the red-orange shrimp give off wastes rich in nitrogen, which are broken down by bacteria and taken up in turn by the algae. It's a question of balance.

The saline environment is chemically buffered and contains a carefully selected mixture of inorganic trace minerals. A key to the success of the system is the inclusion of the coral skeleton, which serves as an inert substrate on which the algae cling and the shrimp graze.

Also living in the colony are large numbers of single-celled creatures and small nematodes that are cultured together and injected before the glass is sealed.

The art in designing such a closed ecological system is in selecting a group of organisms that are relatively evenly matched in the competition for the raw materials of life. Monopoly means death in the EcoSphere.

The first EcoSpheres using this species of shrimp were made 6 years ago and are still going strong, but the system will not last forever.

For some reason the shrimp, with a lifespan between five and 10 years, do not reproduce in the artificial environments.

Although the EcoSphere's occupants don't require feeding, the system has fairly strict temperature and sunlight requirements.

Comparison with the Miracle Whip Microcosm

Neither NASA, JPL, nor SEBRA ever saw my Miracle Whip Microcosm, whose soil, water, and animal and plant inhabitants are scooped up randomly from miscellaneous standing water sources. Those creatures are mostly not even identified, much less scientifically selected and matched to each other.

The wonder of this approach — which is lost in the EcoSphere — is that the jar's contents sort themselves out and balance their ecosystem according to their own individual properties and their suitability to this particular environment, including the presence of each other. This same kind of sorting and self-balancing process has been happening continuously throughout our Earth for the past couple of billion years. In fact, we can thank such processes for our very existence.

The Eco-jar community created by this method is far more richly diverse than the one in the EcoSphere, usually with numerous visible tiny moving creatures, even more than you may have thought because they were clinging to the green stuff or submerged in the mud when you first put them into the jar.

References

1. Immense quantities of living bacteria, fungal spores, and plant material have been found in the clouds, serving as the nuclei of ice crystals and water droplets. Nancy Atkinson, "First Observations of Biological Particles in High-Altitude Clouds." Universe Today (universetoday.com), May 18, 2009.

2. Susan H. Frohling, Chief Trademark Counsel, Kraft Foods. Personal correspondence, September 8, 2010.

3. Ronald Wright, "A Short History of Progress." House of Anansi Press, Toronto, 2004.

4. The first lesson from the 2010 BP oil spill in the Gulf of Mexico should be that, regardless of the apparent safety record of other deep-water drilling, and the difficulty of confirming or refuting reports of multiple other smaller leaks on the ocean floor, there is currently no existing technology that can reliably deal with serious unforeseen events a mile under water.

5. Edwin Good, "Genesis 1-11: Tales of the Earliest World." Stanford University Press, Stanford, CA, 2011. A new literal translation from original Hebrew by a former Professor of Biblical Studies at Stanford University.

6. Harold Morowitz, "The Emergence of Everything: How the World Became Complex." Oxford University Press, 2004.

7. Michael Chabon, "The Omega Glory." Details Magazine, Condé Nast, January 2006.

8. Robert Kunzig, "Climate Milestone: Earth's CO_2 Level Passes 400 ppm." National Geographic News, May 9, 2013.

9. Jerry Silver, "Global Warming and Climate Change Demystified." McGraw-Hill, 2008.

10. Union of Concerned Scientists, personal correspondence.

11. HRH Charles, Prince of Wales, "Harmony: A new way of looking at our world." Blue Door/Harper Collins, 2010.

12. Donnella Meadows, "Thinking in Systems: A Primer." Chelsea Green Publishing, December 2008.

13. Walter Cannon, "The Wisdom of the Body." W.W. Norton, N.Y., 1932.

14. This kind of homeostatic feedback is often referred to as "negative" feedback, but I have chosen to give it the more positive name "homeostatic."

15. "Greed" is a more ambiguous word than the others. Webster defines it as "an inordinate or despicable degree of acquisitiveness," but we have no objective yardstick for "inordinate" or "despicable." Presumably it means "a lot more than most people, and certainly more than I have."

16. I knew a college teacher of "social science" who said he spent the first half of each semester trying to convince the students that there is no such thing as social science.

17. George Soros, "Open Society: Reforming Global Capitalism." Public Affairs, New York, 2000.

18. Herman Daly and Joshua Farley, "Ecological Economics: Principles And Applications." Island Press, 2004.

19. Herman Daly, ed., "Ecology, economics, ethics: Essays toward a steady state economy." W.H. Freeman & Co., October 1980.

20. It is important to note that these meanings and values are not intrinsic to the ingredients themselves.

21. For current and practical purposes, "lost" does not mean complete loss. On a geological time scale, what is now the bottom of the sea will quite possibly become dry land, and vice versa, in the future. Sadly, such long-term upheavals cannot be portrayed in the Eco-jar within a lifetime.

22. Joseph S. Davis, "Essays in the Earlier History of American Corporations." Harvard University Press, 1917.

23. Cited by Paul Cienfuegos in an essay on paulcienfuegos.com.

24. John Kenneth Galbraith, "The New Industrial State." Houghton Mifflin Company, Boston, 1967.

25. Bill Merrill, "Wisdom of the Tools: What Directs the World's Work?" Homeostasis Press, 2006.

26. Vandana Shiva, "Biopiracy: Plunder of Nature and Knowledge." South End Press, July 1999.

27. My main source of information on this subject, other than Shiva's writings, has been from "Breton Woods" and associated searches on the internet.

28. William Ophuls, "Plato's Revenge: Politics in the Age of Ecology." Massachusetts Institute of Technology, 2011.

29. Wade Davis, "The Wayfinders: Why Ancient Wisdom Matters in the Modern World." House of Anansi Press, Toronto, October 2009.

30. Winona LaDuke is an American Indian activist, environmentalist, economist, and writer of Anishinaabekwe (Ojibwe) descent.

31. Debra Harry, an international advocate for indigenous wisdom and culture, is on the staff of The Indigenous Peoples Council on Biocolonialism (ipcb.org).

32. For example, Wade Davis, Ian Mackenzie, and Shane Kennedy, "Nomads of the Dawn: The Penan of the Borneo Rain Forest." Pomegranate Communications, Portland OR, 1995.

33. "Monsanto wins lawsuit over pollen." The Oregonian, Portland, March 30, 2001.

34. Vandana Shiva, "Stolen Harvest: The Hijacking of the Global Food Supply." South End Press, 2000.

35. Garrett Hardin, "The Tragedy of the Commons." Science, December 13, 1968.

36. James Olson, "The New Oil: Should private companies control our most precious natural resource?" Newsweek, October 8, 2010.

37. Pickens' unmapped aquifer seems as though it should qualify as a part of the commons.

38. Earl Cook, "Perspectives on Needs and Supplies of Resources." Science, Vol. 191, Feb. 1976.

39. Greg Helten, "Canada May Lift Moratorium on West Coast Oil, Gas Drilling." Environmental News Service, May 17, 2004.

40. Manek Dubash. Moore's Law is dead, says Gordon Moore. Techworld, 13 April 2005.

41. Wes Jackson and Wendell Berry. Interview, "The Sun" magazine, October 2010.

42. Herman Daly and Kenneth Townsend, eds., "Valuing the Earth: Ecology, Economics, Ethics." The MIT Press, January 1993.

43. Thomas Hobbes, "Leviathan" (Chapter XIII). 1660.

44. Alfred, Lord Tennyson, "In Memoriam (The Way of the Soul)". 1849.

45. H.O. Lancaster, "Expectations of Life." Springer, 1990.

46. Aristotle, "Physics," Book II. 192 B.C.E.

47. Planet Drum Foundation, "Reinhabiting Cities and Towns: Designing for Sustainability" and "Food as place: Bioregional agriculture." San Francisco CA, Winter 1993/1994.

48. Ivan Illich, "Energy and Equity." Le Monde, 1973.

CPSIA information can be obtained at www.ICGtesting.com
Printed in the USA
LVOW02s0355180214

374142LV00001B/1/P